HOW TO THINK ABOUT
ABSTRACT ALGEBRA

HOW TO THINK ABOUT
ABSTRACT ALGEBRA

LARA ALCOCK

Mathematics Education Centre, Loughborough University

OXFORD
UNIVERSITY PRESS

Great Clarendon Street, Oxford, OX2 6DP,
United Kingdom

Oxford University Press is a department of the University of Oxford.
It furthers the University's objective of excellence in research, scholarship,
and education by publishing worldwide. Oxford is a registered trade mark of
Oxford University Press in the UK and in certain other countries

© Lara Alcock 2021

The moral rights of the author have been asserted

First Edition published in 2021

Published in the United States of America by Oxford University Press
198 Madison Avenue, New York, NY 10016, United States of America

British Library Cataloguing in Publication Data

Data available

Library of Congress Control Number: 2020946044

ISBN 978-0-19-884338-2

Printed and bound in the UK
by CPI Group (UK) Ltd, Croydon, CR0 4YY

PREFACE

This preface is written primarily for mathematics lecturers, but students might find it interesting too. It describes differences between this book and other Abstract Algebra texts and explains the reasons for those differences.

his book is not like other Abstract Algebra[1] books. It is not a textbook containing standard content. Rather, it is designed as pre-reading or concurrent reading for an Abstract Algebra course. I do mean that it is designed for *reading*, which is important because students are often unaccustomed to learning mathematics from books, and because research shows that many do not read effectively. This book is therefore less dense and more accessible than typical undergraduate texts. It contains serious discussions of central Abstract Algebra concepts, but these begin where the student is likely to be. They make links to earlier mathematics, refute common misconceptions, and explain how definitions and theorems capture intuitive ideas in mathematically sophisticated ways. The narrative thus unfolds in what I hope is a natural and engaging style, while developing the rigour appropriate for undergraduate study.

Because of this aim, the book is structured differently from other texts. Part 1 contains four chapters that discuss not the content of Abstract Algebra but its structure, explaining what it means to have a coherent mathematical theory and what it takes to understand one. There is no 'preliminaries' chapter; instead, notations and definitions are introduced where they are first needed, meaning that they are spread across the text (though a symbol list is provided on pages xiii–xiv). This means that a

---

[1] 'Abstract Algebra' should probably not have upper-case 'A's, but I want to make the subject name distinct from other uses of related terms.

student reading for review might need to use the index more than usual, so the index is extensive.

A second difference is that not all content is covered at the same depth. The five main chapters in Part 2 each contain extensive treatment of their central definition(s), especially where students are known to struggle. They also discuss selected theorems and proofs, some of which are used to highlight strategies and skills that might be useful elsewhere, and some of which are used to draw out structural similarities and explain theory development. But these chapters aim to prepare a student to learn from a standard Abstract Algebra course, rather than to cover its entire content.

A third difference is that the order of the content is relatively unconstrained by logical theory development. Theory is explicitly discussed: Part 1 provides information on the roles of axioms, definitions, theorems and proofs, and Part 2 encourages attention to logical argument. But numerous sections provide examples before inviting generalization, or introduce technical terms informally before they are defined, or observe phenomena to be formalized later. I take this approach with care, highlighting informality and noting where formal versions can be found. But I consider it useful because the central ideas of early Abstract Algebra are so tightly interwoven—really I would like to introduce all of the book's main ideas simultaneously. Obviously that is impossible, but I do want to keep some pace in the narrative, to prioritize concepts and important relationships over technicalities. I realize, of course, that this approach goes against the grain of mathematical presentation and means that this is not a book that a typical course could 'follow'. But I am content with that—the world is full of standard textbooks developing theory from the bottom up, and my aim is to provide an alternative with a focus on conceptual understanding.

Finally, this book explicitly discusses how students might make sense of Abstract Algebra as it is presented in lectures and in other books. I realize that this, too, is contentious: many mathematics lecturers place high value on constructing ideas and arguments, and some have worked hard to develop inquiry-based Abstract Algebra courses. I am all for inquiry-based courses, and for any well-thought-out approach that allows students to reinvent mathematical ideas through independent or collaborative problem solving. But the reality is that most mathematics lectures are still just that: lectures. Many lecturers are constrained by class sizes well into the hundreds, and flipped classroom models might promote student engagement but it is not obvious how to use them effectively to develop the theory of Abstract Algebra. Because of this, and because few students follow every detail of their lectures, an important student task is to make sense of written mathematics. Research shows that the typical student is capable of this but ill-informed regarding how to go about it. This book tackles that problem—it aims to deliver students who do not yet know much Abstract Algebra but who are ready to learn.

A book like this would be impossible without work by numerous researchers in mathematics education and psychology. In particular, the self-explanation training in Chapter 3 was developed in collaboration with Mark Hodds and Matthew Inglis (see Hodds, Alcock & Inglis, 2014) on the basis of earlier research on academic reading by authors including Ainsworth and Burcham (2007), Bielaczyc, Pirolli, and Brown (1995), and Chi, de Leeuw, Chiu and LaVancher (1994). More information on the studies we conducted can be found in Alcock, Hodds, Roy and Inglis (2015), an article in the *Notices of the American Mathematical Society*. The bibliography contains extensive references to decades' worth of research on student learning about specific concepts in Abstract Algebra and about general proof-based mathematics. I encourage interested readers to investigate further.

My specific thanks go to Ant Edwards, Tim Fukawa-Connelly, Kevin Houston, Artie Prendergast-Smith, Adrian Simpson, Keith Weber, and Iro Xenidou-Dervou, all of whom gave valuable feedback on chapter drafts. Similarly, to careful readers Romain Lambert, Neil Pratt, and Simon Goss, who were kind enough to point out errors. I am particularly indebted to Colin Foster, who read a draft of the entire book before its chapters went to anyone else. I am also grateful as ever to the team at Oxford University Press, including Dan Taber, Katherine Ward, Chandrakala Chandrasekaran and Richard Hutchinson. Finally, this book is dedicated to Kristian Alcock, who has never known me not to be writing it. I think he will be glad and amazed to see it in print.

CONTENTS

Part 2 Topics in Abstract Algebra

SYMBOLS

Symbol	Meaning	Section
$\Rightarrow$	(which) implies (that)	1.2
$\in$	in or (which) is an element of	1.3
$\mathbb{N}$	the set of all natural numbers	1.3
ρ	the Greek letter 'rho'	1.3
$\exists$	there exists	2.1
$\mathbb{Z}$	the set of all integers	2.1
$\{ah\|h \in H\}$	the set of all ah such that h is in H	2.1
$\forall$	for all	2.2
$*$	'star' (denotes a binary operation)	2.2
$\mathbb{Q}$	the set of all rational numbers	2.2
$\mathbb{R}$	the set of all real numbers	2.2
$\mathbb{C}$	the set of all complex numbers	2.2
$\notin$	not in or (which) is not an element of	2.2
$\mathbb{Z}\backslash\{0\}$	the set of all integers excluding zero	2.2
$3\mathbb{Z}$	$\{3n\|n \in \mathbb{Z}\}$	2.2
$\circ$	function composition	2.5
$\mathbb{Z}_{12}$	the set, group or ring of congruence classes modulo 12	3.3
$+_{12}$	addition modulo 12	3.3
$14 \equiv 2\,(\mathrm{mod}\,12)$	14 is congruent to 2 modulo 12	3.3
$\Leftrightarrow$	if and only if or (which) is equivalent to	3.3
$a + 12\mathbb{Z}$	$\{a + 12x\|x \in \mathbb{Z}\}$	3.4
$x \sim y$	x is related to y	3.4
$X \cap Y$	the intersection of sets X and Y	3.4
$\emptyset$	the empty set	3.4
$X \subseteq Y$	X is a subset of Y	3.4
$X \times X$	the set of pairs (x, y) where $x, y \in X$	3.4
$C^0(\mathbb{R}, \mathbb{R})$	the set of continuous functions from $\mathbb{R}$ to $\mathbb{R}$	5.4
$M_{2\times2}(\mathbb{R})$	the set of 2×2 matrices with entries in $\mathbb{R}$	5.5
$(G, *)$	the group with set G and operation $*$	6.1
$\langle g \rangle$	the set generated by g	6.3
$\|G\|$	the order of the group G	6.3

Symbol	Meaning	Section
$\langle a \vert a^{12} = e \rangle$	the group generated by a such that $a^{12} = e$	6.4
U_n	the set of nth roots of unity	6.6
$\cong$	is isomorphic to	6.6
U	the unit circle	6.6
$GL(n, \mathbb{R})$	the general linear group of degree n over $\mathbb{R}$	6.6
V	the Klein four-group	6.8
S_n	the symmetric group of degree n	6.9
$\vert G{:}H \vert$	the index of the subgroup H in the group G	7.7
$X \cup Y$	the union of sets X and Y	7.7
ϕ	the Greek letter 'phi'	8.1
ψ	the Greek letter 'psi'	8.1
$\mathbb{Z}[x]$	the ring of polynomials with coefficients in $\mathbb{Z}$	9.2

INTRODUCTION

This short introduction discusses the place of Abstract Algebra in typical undergraduate degree programmes and the challenges it presents. It then explains this book's content and intent.

Mathematics students typically encounter Abstract Algebra as one of their first theorems-and-proofs courses. For those in UK-like systems, where people specialize early, it might be taught at the beginning of a mathematics degree, or perhaps in the second term or second year. For those in US-like systems, where people specialize late, it will more likely be an upper-level course for mathematics majors in their junior or senior year. Either way, a course might not actually be called *Abstract Algebra*. That name is an umbrella term for the theory of groups, rings, fields and related structures, so a first course might be titled *Group Theory* or *Groups, Rings and Fields* or something more basic like *Sets and Groups*.

Whatever the precise arrangements, Abstract Algebra is usually studied in parallel with other subjects such as Analysis. Both Abstract Algebra and Analysis involve a shift in mathematical emphasis: students who think of mathematics as a set of algorithms must now learn to focus on definitions, theorems and proofs. Part 1 of this book provides advice on that. But the two subjects can feel quite different, and awareness of the differences might help with understanding their respective challenges.

In Analysis, the main challenge is the reasoning: the subject makes heavy use of logically complex statements, which few students are well equipped to process. But its objects are relatively graspable: numbers, sequences, series and functions are already familiar or readily represented in diagrams. In Abstract Algebra, the main challenge is almost the opposite. The logic is more straightforward, but many of the objects are less familiar and less easy to represent. For students, this can render the subject somewhat meaningless—even those who do well might not develop a strong sense of what it is 'about'. That was my experience. Despite having a very good lecturer,[1] I never really liked Abstract Algebra because I never really *got* it. I could perform the manipulations, apply the theorems, and reconstruct the proofs, but I didn't really understand what it all meant.

I now believe that this happened for two reasons. The first is general but particularly relevant in Abstract Algebra: mathematics is hierarchical, with each level building on the last. Shifting up a level often requires compressing some aspect of your understanding in order to to think of it in relation to four or five new things. If you haven't compressed it enough, this is difficult, and higher levels can seem like meaningless symbol-pushing. Abstract Algebra involves a lot of compression, and this book will point out explicitly where it is needed.

The second reason is that I didn't access good representations for Abstract Algebra's key ideas. I like visual representations—images that enable me to 'see' how concepts are related and to develop intuition for why things work as they do. I don't often get that feeling from algebraic arguments, no matter how sure I am that each step is valid. And Abstract Algebra didn't seem to have many visual representations, so I didn't find much to hold on to. I don't blame my lecturer for this—I failed rather badly to keep up, so I didn't follow his lectures effectively and I probably missed some enlightening explanations. But intuition for the subject can be developed using diagrams and tables, so this book contains many of those.

This book also contains explicit discussion of what Abstract Algebra *is*, beginning in Chapter 1 with sections on what is abstract about Abstract Algebra, and what is algebraic about it. The remainder of Part 1 discusses axioms and definitions and their roles in mathematical theory (Chapter 2), theorems and proofs and productive ways to interact with these (Chapter 3), and research-based strategies for effective learning (Chapter 4). Please read Chapter 4 even if you are a successful

---

[1] In the UK we say 'lecturer' where those in US-like systems might say 'instructor' or 'professor'.

student—you might find that you can tweak your strategies to improve your learning or reduce your workload or both.[2]

Part 2 covers topics in Abstract Algebra, starting with binary operations (Chapter 5) and moving on to groups and subgroups (Chapter 6), quotient groups (Chapter 7), isomorphisms and homomorphisms (Chapter 8) and rings (Chapter 9). Because this is not a standard textbook, it does not try to 'cover' all of the relevant content for these topics. Instead, it treats the main ideas in depth, using examples and visual representations to explain ways to think about them accurately. Each chapter also includes selected theorems and proofs and discusses relationships between topics.

Because Abstract Algebra is a tightly interconnected theory, early chapters often touch on ideas not formalized until later. For this reason, I recommend reading the whole book in order, although each chapter should also be readable as a self-contained unit. If you get stuck, remember that there is an extensive index, and that pages xiii–xiv list where symbols are explained in the text. If it is practical, I also recommend reading the entire book before starting an Abstract Algebra course. I intend to set you up with meaningful understanding of the main ideas and a good grasp of how to learn effectively, so you will likely get maximum benefit by reading before you start. However, if you have come to this book because your course has begun and you find yourself lost in the abstractions, it should provide opportunities to rework your understanding so that you can engage effectively. Either way, I hope that Abstract Algebra deepens your understanding of both familiar mathematics and higher-level theory.

---

[2] Compared with *How to Study for a Mathematics Degree* and its American counterpart *How to Study as a Mathematics Major*, the advice in Chapter 4 is condensed, specific to Abstract Algebra, and more explicitly linked to research on learning.

PART I
Studying Abstract Algebra

CHAPTER 1

What is Abstract Algebra?

This chapter contrasts Abstract Algebra with the algebra studied in earlier mathematics. It highlights the subject's focus on validity of algebraic manipulations across a range of structures. It then describes three approaches common in Abstract Algebra courses: a formal approach, an equation-solving approach and a geometric approach.

1.1 What is abstract about Abstract Algebra?

Abstract Algebra is abstract in the same sense in which other human thinking is abstract: its concepts can be instantiated in multiple ways. For instance, you recognize things like trees and windows. You can do that because you understand the abstract ideas 'tree' and 'window', and you can match them to objects in the world. You do not need to look at one tree to identify another; you do it by reference to the abstract idea.

Now, trees and windows are physical objects—you can walk up and touch them, and identify them by sight. But you can think about concepts that are more abstract, too. For instance, you can identify an aunt. You do that by reference to a criterion: is this a female person with a sibling who has children? If yes, it's an aunt. If no, it's not. Moreover, you can think about abstract concepts that are not single objects, like *family*. A family includes multiple people—perhaps many—who are related to one another—genetically or by marriage or by other caring relationships. Families vary a lot, and you couldn't necessarily recognize a family by

sight or by checking simple criteria. But you nevertheless understand the idea.

Abstract Algebra is about concepts that are somewhat like each of these more abstract ideas. They are like aunts in that they are defined by criteria. Abstract Algebra is stricter, though. Everyday human concepts, even defined ones like 'aunt', tend to be used flexibly. Is your mum's brother's female partner your aunt? Maybe, maybe not. And where I grew up, adult female neighbours and friends were commonly referred to as 'Auntie', even where there were no family relationships. Such flexibility doesn't happen in Abstract Algebra, because mathematical concepts are specified by precise definitions about which all mathematicians agree.[1]

The concepts of Abstract Algebra are like families in that not all are just single things: some are sets with particular internal relationships. This means that they can be big and complex, though that does not necessarily make them hard to think about. A single tree, after all, might have tens of branches and thousands of leaves—it has lots of internal structure, but you can treat it as a single thing. Mathematical objects can be like this too. For instance, *the set of all even numbers* is infinite and has lots of internal structure, but again you can treat it as a single thing. Such thinking is important in Abstract Algebra: often it is useful to switch between examining an object's internal structure and thinking of it as a unified whole.

1.2 What is algebraic about Abstract Algebra?

To understand what is algebraic about Abstract Algebra, it is probably useful to consider what is algebraic about earlier algebra. Many students think of algebra as something you *do*, where *doing algebra* means manipulating an expression or equation in valid ways to arrive at another. This is often in service of a goal: solving a mechanics problem, say. And thinking of algebra in this way is not wrong—certainly it captures most students'

---

[1] More accurately, mathematicians agree about the principle that concepts should be defined in this way, and most undergraduate mathematics works like this. But, historically, there were debates about how best to define everything, and such debates continue in developing subjects.

experience prior to undergraduate mathematics. But it is not enough to grasp the aims of Abstract Algebra.

Abstract Algebra focuses not on performing algebraic manipulations but on understanding the mathematical structures that make those manipulations valid. To see what I mean, consider this algebraic argument (the arrow '$\Rightarrow$' means 'implies').

$$x(x + y) = yx$$
$$\Rightarrow x^2 + xy - yx = 0$$
$$\Rightarrow \qquad\qquad x^2 = 0$$
$$\Rightarrow \qquad\qquad x = 0.$$

Probably you can write such arguments quickly, fluently and with only occasional errors. But why exactly is each step valid? One step (which?) assumes that $xy = yx$. This is valid because multiplication is *commutative*, meaning that xy and yx always take the same value. Another step (which?) assumes that if $x^2 = 0$ then $x = 0$. This is valid because 0 is the only number that, when squared, gives 0. But such assumptions rely on properties of operations and objects. Multiplication is commutative, but not all operations share this property. Division is not commutative, for instance: x/y could not be replaced by y/x. And not all objects behave like numbers. If x and y were 2×2 matrices,[2] we could not assume that $xy = yx$ because matrix multiplication works like this:

$$\begin{pmatrix} x_{11} & x_{12} \\ x_{21} & x_{22} \end{pmatrix} \begin{pmatrix} y_{11} & y_{12} \\ y_{21} & y_{22} \end{pmatrix} = \begin{pmatrix} x_{11}y_{11} + x_{12}y_{21} & x_{11}y_{12} + x_{12}y_{22} \\ x_{21}y_{11} + x_{22}y_{21} & x_{21}y_{12} + x_{22}y_{22} \end{pmatrix}.$$

So, for example,

$$\begin{pmatrix} 1 & 2 \\ 3 & 4 \end{pmatrix} \begin{pmatrix} 5 & 6 \\ 7 & 8 \end{pmatrix} = \begin{pmatrix} 5 + 14 & 6 + 16 \\ 15 + 28 & 18 + 32 \end{pmatrix} = \begin{pmatrix} 19 & 22 \\ 43 & 50 \end{pmatrix}$$

---

[2] This book includes examples based on matrices and complex numbers. If you are studying in a UK-like system and have not come across these, you can find introductions in A-level Further Mathematics textbooks or reliable online resources.

but

$$\begin{pmatrix} 5 & 6 \\ 7 & 8 \end{pmatrix} \begin{pmatrix} 1 & 2 \\ 3 & 4 \end{pmatrix} = \begin{pmatrix} 5+18 & 10+24 \\ 7+24 & 14+32 \end{pmatrix} = \begin{pmatrix} 23 & 34 \\ 31 & 46 \end{pmatrix}.$$

Similarly, we could not assume that if $x^2 = 0$ then $x = 0$. The matrix

$$\begin{pmatrix} 0 & 1 \\ 0 & 0 \end{pmatrix}$$

is not the zero matrix, but nevertheless

$$\begin{pmatrix} 0 & 1 \\ 0 & 0 \end{pmatrix} \begin{pmatrix} 0 & 1 \\ 0 & 0 \end{pmatrix} = \begin{pmatrix} 0+0 & 0+0 \\ 0+0 & 0+0 \end{pmatrix} = \begin{pmatrix} 0 & 0 \\ 0 & 0 \end{pmatrix}.$$

Algebraic validity therefore depends upon properties of both *binary operations*, including multiplication, division and others to be discussed in Chapter 5, and the *sets* on which these operate, which might be sets of numbers, matrices or objects of other types—again, see Chapter 5. Binary operations might be commutative on some sets but not on others. In some sets there are few ways to combine objects to give zero; in others, there are many. Thus, 'facts' that are true for one operation on one set do not necessarily hold elsewhere, and Abstract Algebra requires concentration to ensure that you do not overgeneralize from a familiar context. I won't lie: this is hard. When you are accustomed to 'doing' algebra in numerical contexts, it might not require much effort. The manipulations become natural enough that you do them easily, much as you might walk or type easily. And concentrating on something that you do easily feels weird and disruptive. If you concentrate on your muscles as you walk, you become slow and ungainly. If you can touch-type, but you force yourself to look at the keyboard and think about which letters you want, you might find that you can't type at all. Focusing on why algebraic manipulations are valid can feel similar: slow, clunky and therefore like a step backward in your learning rather than a step forward. It takes discipline, and for a while might feel frustrating.

But the reward is worth it. Differences occur in the detail, but at larger scales Abstract Algebra reveals strikingly similar structures. Sets and operations that appear quite different turn out to have numerous common properties. This sort of thing excites pure mathematicians. But even if you're not by nature one of those—if you're into mathematics for its practical applications—I encourage you to be open to the ideas. There is pleasure in recognizing structures that cut across the subject.

1.3 Approaches to Abstract Algebra

To appreciate Abstract Algebra, you will need to engage effectively with your course. And courses differ. Lecturers have individual approaches, and even those who follow books will emphasize some things and skip over others. That said, many Abstract Algebra courses start with group theory. Some start with rings or a combination of structures but, as the group theory beginning is common, it forms the greater part of this book.

In teaching group theory, there are three broad approaches: a *formal* approach, an *equation-solving* approach and a *geometric* approach. Your course will likely have something in common with at least one of these, and this book will draw on all three. Here I will describe each in turn, commenting on their advantages and disadvantages. The descriptions are necessarily caricatures, but they give a flavour of what you might encounter.

A *formal* approach is in one sense the simplest. Lecturers taking this approach tend to begin with definitions, like that for *group* shown below.[3] This is explained in detail in Chapters 2, 5 and 6; for now, you might like to know that the symbol '$\in$' means '(which) is an element of' and is often read simply as 'in'.

Definition: A **group** is a set G with a binary operation $*$ such that:

Closure for every $g_1, g_2 \in G$, $g_1 * g_2 \in G$;

Associativity for every $g_1, g_2, g_3 \in G$, $(g_1 * g_2) * g_3 = g_1 * (g_2 * g_3)$;

Identity there exists $e \in G$ such that for every $g \in G$, $e * g = g * e = g$;

Inverses for every $g \in G$, $\exists g' \in G$ such that $g * g' = g' * g = e$.

---

[3] If your course uses a definition that omits the *closure* criterion, see Section 5.7.

After introducing a definition, lecturers taking a formal approach will use it to prove general theorems. So a formal approach foregrounds the key concepts and deductive nature of advanced mathematics. And some people like it. It's tidy, it's slick and it leads to a concise set of lecture notes. It avoids expansive discussion of ways in which abstract theory can be instantiated, so there is minimal distraction and it should be clear that definitions, theorems and proofs are central. It also saves time, so a course can get to deeper results quite quickly. Some people find that motivating because deeper results are more interesting.

The speed of a formal approach can, however, present challenges. A theory presented with few examples provides little opportunity to build intuition, so that students might learn many theorems but find it difficult to decide which to use in problem solving. It is also, for some, actively *de*motivating. If your lecturer takes this approach but you value intuition, you might can find yourself staring at definitions, theorems and proofs and thinking, 'Well, that seems to be valid, but why should I care? Why define a group that way in the first place?'

The *equation-solving* approach deals with this problem. A lecturer taking this approach might start by inviting students to consider what properties are needed to ensure that equations of the form $x + a = b$ always have solutions. Working this out involves solving this simple equation while attending to those properties. We want to subtract a from both sides, writing

$$x + a = b$$
$$\Rightarrow \quad x = b - a.$$

What assumptions does that require? First, subtraction must be possible; a must have an *additive inverse* $-a$. This cannot be taken for granted: in the set of natural numbers $\mathbb{N} = \{1, 2, 3, \ldots\}$, the elements have no additive inverses. Further assumptions are revealed by unpacking what we are doing.

$$x + a = b$$
$$\Rightarrow (x + a) + (-a) = b + (-a) \quad \text{(adding } -a \text{ to both sides)}$$
$$\Rightarrow x + (a + (-a)) = b + (-a) \quad \text{(reordering operations on the left)}$$
$$\Rightarrow \quad x + 0 = b + (-a) \quad \text{(using } a + (-a) = 0\text{)}$$
$$\Rightarrow \quad x = b + (-a) \quad \text{(using } x + 0 = x\text{)}.$$

You see what I mean about the challenge of concentrating on validity. If you are wondering why anyone would bother, check your assumptions about sets and operations. Would equivalent intermediate steps be valid if the operation were multiplication, so that the original equation read $ax = b$? Would the addition and multiplication versions work if the objects were matrices? If not, what could go wrong?

Then consider the intermediate steps. For the first, as noted above, the object a must have an *additive inverse* $-a$. For the second, the operations are reordered so that $a + (-a)$ is calculated before $x + a$; for this, the operation must be *associative*. The third uses the inverse property directly. The fourth uses the fact that 0 is the additive *identity* element. Finally, the result $b + (-a)$ must be in the set; the set must be *closed under the operation*. Now, did you notice what this means? Guaranteeing solvability for equations like $x + a = b$ or $ax = b$ requires closure, associativity, an identity element and inverses. In other words, it requires a structure that satisfies the definition of *group*. That definition does not come from nowhere.

The equation-solving approach thus links Abstract Algebra to students' earlier algebraic experience, so it provides a natural way in. And different equation types demand different structures, leading to theory about rings, fields and so on. But this approach alone gives little sense of group structures that exist beyond familiar sets and operations.

The geometric approach addresses that. It starts with structures—and indeed objects—that are, for many students, unfamiliar. The objects are the *symmetries* of shapes such as equilateral triangles and squares, where each symmetry is a transformation after which the shape 'looks the same'. For instance, an equilateral triangle has the six symmetries represented below (the spots are just to keep track). These comprise an identity e ('leave the triangle where it is'), two rotations and three reflections. With the first rotation denoted by ρ (the Greek letter 'rho'), why does it make sense to denote the second by ρ^2? Why do you think the identity is included?

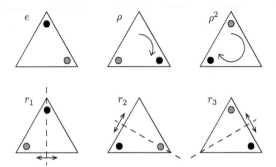

This use of the word *symmetry* differs from colloquially saying that the triangle 'is symmetrical'. It is more precise in that the six symmetries are recognized as distinct. Moreover, 'symmetry' is a noun (as in 'a reflection', 'a rotation'). This is important because these symmetries can be treated as elements of a set. Indeed, they form a group under the binary operation of *composition*, where composing two symmetries means performing one then the other. For instance, performing ρ then r_1 has the same effect as performing the single symmetry r_2.

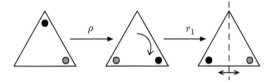

What is required to establish that these symmetries form a group under composition? Which properties are easy to check and which are more difficult? Symmetry groups, including alternative notations, will be discussed in Sections 5.6 and 6.7–6.8.

The geometric approach to Abstract Algebra is often favoured by lecturers who implement inquiry-based learning,[4] because students can manipulate physical triangles to develop intuition and explore conjectures.

---

[4] Instead of lecturing at the board, the lecturer sets problems for students to work on together.

You might enjoy this if—like me—you like visual representations. I can imagine the objects, although it takes concentration to remember that the group elements are *symmetries*—rotations and reflections—not the triangle or its corners (see Section 6.7). If you are more of a formalist, you might think that manipulating shapes is a bit childish, or that it wastes time that could be used for real theory. That's fair enough, though I encourage you to embrace the intuition that can come from such activity.

One broader disadvantage of a geometric approach is that it privileges one type of structure, and does so in a vivid, attention-grabbing way. This can give the impression that the whole of group theory is about symmetry groups, or at least that these are somehow the most important. It might be less obvious that group theory also applies to familiar structures, and students might therefore miss opportunities to connect their mathematical knowledge. So, while visual representations are useful, I encourage you to develop your understanding of a range of groups, so that you can use different insights and appreciate the scope of the theory. This book is designed to help you do that.

CHAPTER 2

Axioms and Definitions

This chapter explains how axioms and definitions fit into mathematical theories, and describes ways to relate them to examples. It notes relationships between axioms, definitions and the key concepts of group and ring, and introduces ideas to be formalized in Part 2. It concludes by discussing object types in Abstract Algebra and corresponding notations.

2.1 Mathematical axioms and definitions

Axioms and definitions form the basis of any mathematical theory, where a *theory* is an interlinked network of concepts and results. Axioms and definitions are the assumptions and agreements from which mathematicians build a theory by proving *theorems*, where a theorem is a true statement about one or more mathematical concepts. This is represented in the diagram below.

Understanding axioms and definitions is therefore crucial. And it is important to know what this involves, because mathematical concepts are not like everyday concepts. That might be obvious for axioms—in everyday life, no one really talks about those. People do talk about definitions, but everyday definitions are *not the same* as mathematical definitions. In fact, everyday definitions are mostly ignored because we usually learn about concepts by exposure to examples. If we encounter an unfamiliar concept, we might look up its definition, but we might infer its meaning from the context. People do this imperfectly, but that is rarely disastrous. We usually grasp roughly the intended meaning, and many

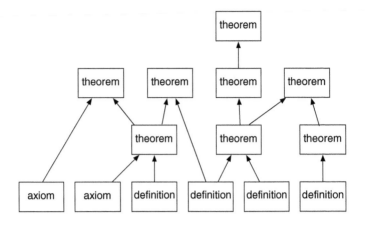

concepts in any case have 'woolly edges' around which people do not quite agree.

Mathematical definitions are not like this. Mathematical definitions specify properties, and the words they define are used exactly and only for things that satisfy those properties. For example, a definition of *even number* appears below. I recommend reading such definitions—and all mathematics—aloud: it is important to own the language so that you can use it naturally when reading and when communicating with others. Here, '∃' means 'there exists', '$\mathbb{Z}$' denotes the set of all *integers* (whole numbers, including negatives and zero) and '∈' can be read as 'in'.

Definition: A number n is **even** if and only if $\exists k \in \mathbb{Z}$ such that $n = 2k$.

This definition splits numbers cleanly into those that are even and those that are not. There is no ambiguity: $4, -38$ and 0 are even; 5 and -351 are not; 12.001 might be 'close to' the even number 12, but it is not even, and it is certainly not considered even by some people but not by others. This is due to the logic captured in the phrase 'if and only if', which is sometimes easier to process by considering each part separately:

A number n is even *if* $\exists k \in \mathbb{Z}$ such that $n = 2k$;
A number n is even *only if* $\exists k \in \mathbb{Z}$ such that $n = 2k$.

The first part means that a number is even if there exists an appropriate k. The second means that a number cannot be even otherwise—try reading each part aloud with emphasis on the italics. Then, to make the logic clearer, consider the two statements below. These are similarly structured, but something important must be different because one is true and the other is false. Which is which?

A number n is divisible by 2 *if* it is divisible by 4.

A number n is divisible by 2 *only if* it is divisible by 4.

The true statement splits numbers into three partially nested sets because some numbers are divisible by 2 but not by 4. Consider how the logic relates to the diagram.

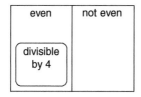

In mathematics, it often happens that one of an *if* and an *only if* is true and the other is false—Chapter 3 discusses theorems with this feature. But definitions all have *if and only if* structures. Unfortunately, that is not always apparent from teaching materials. First, some definitions sound more natural when formulated differently—I will point out where that occurs in this book. Second, mathematicians understand the logic of definitions without having to think about it, so some write 'if' knowing that they really mean 'if and only if'. I think that this overlooks opportunities to help students focus on logic. On the other hand, undergraduates need to learn their subject's norms, to interpret this kind of thing correctly. Whatever your lecturer's approach, bear this information in mind.

Now, the principle behind mathematical definitions is easy enough to work with for even numbers, where the definition is simple and the relevant objects are familiar. It can be harder when a definition is complex and the relevant objects are unfamiliar. Consider the definition below, for instance.

Definition: Suppose that H is a subgroup of G. Then the **left coset** of H containing a is $aH = \{ah | h \in H\}$.

One thing to check is that you can read this aloud. For instance, the expression $aH = \{ah | h \in H\}$ could be read as

'a-big-H is the set of all elements of the form a-little-h, where little-h is in the subgroup big-H.'

In practice, people might not bother with the detail, just saying

'aH is the set of all elements of the form ah, where h is in H.'

That works provided that everyone is looking at the written formulation, so they know which h or H is which. You might want to distinguish them explicitly, though, because h and H are different types of object, as discussed in Section 2.5.

One thing to observe is that this definition sounds natural without an explicit 'if and only if': it is formulated less as a possible property of pre-existing objects, and more as an instruction, as a way to *calculate* left cosets. Coset calculations will be introduced in Section 2.5 and discussed further in Chapter 7. For now, note that applying this definition requires understanding what a subgroup is, what the notation means, and so on. Using it to prove things about cosets in general requires compressing the idea to think about cosets as objects. Chapter 7 discusses that compression, which takes some work. But at least the word *coset* indicates that there is work to do: most people have never heard it so it is clear that it introduces a new concept.

The principle behind mathematical definitions can be even harder to work with when a definition is complex, the relevant objects are unfamiliar, *and the word sounds like an everyday concept*. For instance, as noted in Chapter 1, a central Abstract Algebra concept is that of a *group*. The group concept has a rather long definition, and you probably have little experience with mathematical groups. But you have a lot of experience with everyday groups, and I'm sure you can see that this causes problems. When you hear the word 'group', it will be easier to think about its everyday meaning than its mathematical meaning. Consequently, many students struggle along with an understanding of groups that is overly influenced by the everyday meaning, and rather imprecise. Unsurprisingly, this is not a good foundation for understanding the related mathematical theory.

Recognizing this problem can help students to address a lot of their own difficulties. Students often know that they do not understand things. They might know, for instance, that they do not understand a theorem about cosets because they do not really know what cosets are. The solution to this problem has a simple first step: *study the definition*. But students new to advanced mathematics often do not think to do that, because it is not how they have learned about concepts in everyday life or in earlier mathematics. Consequently, much of this chapter is about ways to process definitions. But its overarching message is simply that definitions are paramount and should command your attention.

2.2 Relating definitions to examples

One way to understand definitions is to relate them to examples. For instance, the concept of *closure* is defined below, where the symbol '∀' means 'for all'. (I will stop reminding you to read aloud now but I recommend that you continue.)

Definition: The set S is **closed** under the binary operation $*$ if and only if $\forall s_1, s_2 \in S, s_1 * s_2 \in S$.

This definition refers to both a set, denoted by S, and a binary operation, denoted by $*$ ('star'). And applying it can be straightforward. For instance, translating from the general set S and operation $*$ to the specific set $\mathbb{N}$ and operation $+$ gives

$$\forall n_1, n_2 \in \mathbb{N}, \ n_1 + n_2 \in \mathbb{N}.$$

This statement is true, so the natural numbers are closed under addition. How about other sets such as $\mathbb{Z}$ (the integers—whole numbers including negatives and zero), $\mathbb{Q}$ (the rational numbers, those of the form p/q where $p \in \mathbb{Z}$, $q \in \mathbb{Z}$ and $q \neq 0$), $\mathbb{R}$ (the real numbers) and $\mathbb{C}$ (the complex numbers)? Are these closed under addition? How about under multiplication?

Check and you will find that all five are closed under both operations: the sum of two real numbers is a real number, the product of two complex numbers is a complex number, and so on. So it might seem like there isn't much to see here. But other sets and binary operations are more interesting. The integers are not closed under division because it is not true that for every $x_1, x_2 \in \mathbb{Z}$, $x_1/x_2 \in \mathbb{Z}$: for instance, $3, 2 \in \mathbb{Z}$ but $3/2 \notin \mathbb{Z}$ (the symbol '$\notin$' means '(is) not in'). Note that this single *counterexample* is enough to show that $\mathbb{Z}$ is not closed under multiplication because of the quantifier '$\forall$'—when mathematicians say 'for all', they really mean it.

When introducing a definition, lecturers commonly apply it to a couple of examples and ask students to apply it to more. I recommend performing at least mental checks in relation to all the examples you can think of. If that turns out to be easy, you have quickly explored the breadth and limitations of the concept. If it turns out to be difficult, you have discovered either a gap in your understanding or a mathematical subtlety that merits more attention.

Here, for instance, it might be tempting to say simply that a set is or is not closed. But that would be mathematically ambiguous—a set on its own cannot be closed because the definition requires a binary operation. Another subtlety is that zero is an integer, but division by zero is not well defined. This renders the closure question moot for division on $\mathbb{Z}$, because division is not even a binary operation on $\mathbb{Z}$. That might strike you as overly fussy, because only one element is problematic: division is defined on nearly all of $\mathbb{Z}$. Mathematicians recognize this, and sometimes restrict to a set such as $\mathbb{Z}\backslash\{0\}$, which means the integers excluding zero. But that raises another subtlety glossed over so far: must the result of a binary operation be in the original set? If it must, then division is not a binary operation even on $\mathbb{Z}\backslash\{0\}$, again due to counterexamples like $3/2$. Division *is* a binary operation on $\mathbb{Q}\backslash\{0\}$, so we could think of $\mathbb{Z}\backslash\{0\}$ as a subset of that. Does that affect closure? Is $\mathbb{Z}\backslash\{0\}$ as a subset of $\mathbb{Q}\backslash\{0\}$ closed under division? No: $3/2$ is still a counterexample.

Chapter 5 will discuss this subtlety further. It will also consider other sets and binary operations, including matrices under addition and multiplication, and functions, symmetries and permutations under composition. If you are familiar with some of these, consider now which are closed under which operations. If you are not, think about restrictions of familiar sets. For instance, the set of all integer multiples of 3 can be denoted by '$3\mathbb{Z}$' because $3\mathbb{Z} = \{3n | n \in \mathbb{Z}\} = \{\ldots, -6, -3, 0, 3, 6, \ldots\}$. Is $3\mathbb{Z}$ closed under addition, or under multiplication? How about $\{z \in \mathbb{C} | z = 0 + yi\}$, the set of all complex numbers of the form $0 + yi$? Is this set closed under addition or multiplication? Can you construct subsets of familiar sets that have different closure properties, perhaps for different binary operations?

Finally, for mathematical usage of everyday words, it is worth thinking about relationships between mathematical definitions and everyday meanings. The two will differ, but usually the words are well chosen. For instance, I would say informally that 'closed' means that it is impossible to 'get out of' the set by combining two of its elements using the binary operation. So it is like having a closed box and, for me, this use of the word makes sense.

A second important concept is *associativity*.

Definition: The operation $*$ is **associative** on the set S if and only if $\forall s_1, s_2, s_3 \in S, (s_1 * s_2) * s_3 = s_1 * (s_2 * s_3)$.

You can understand this in similar ways. Which operations are associative on which sets? Is there an operation that is associative on a whole set but not on a smaller subset, or vice versa? If so, can you give an example? If not, why not? Associativity will be discussed in Chapter 5, but thinking in advance can ground the ideas.

A third important concept is that of an *identity* element, often denoted by 'e' for the German word *einheit*, though you might also see 'id' or 'ι' or your lecturer's own favourite.

Definition: The element $e \in S$ is the **identity** in S with respect to the binary operation $*$ if and only if $\forall s \in S, e * s = s * e = s$.

This definition distinguishes one element of a set with a special property. How would you describe this property informally? I would say that the identity 'doesn't change anything' when combined with other elements using the binary operation. For instance, the identity in $\mathbb{Z}$ with respect to multiplication is 1, because

$$\forall x \in \mathbb{Z}, 1x = x1 = x.$$

What is the identity in $\mathbb{Z}$ with respect to addition? It is 0, because

$$\forall x \in \mathbb{Z}, 0 + x = x + 0 = x.$$

Thus it makes no sense to discuss 'the' identity in $\mathbb{Z}$; this definition too requires both a set and a binary operation. That said, people do abuse the language when an operation is established. For instance, the 2×2 matrix

$$\begin{pmatrix} 1 & 0 \\ 0 & 1 \end{pmatrix}$$

is often referred to as 'the' identity matrix. Which is it, a multiplicative or an additive identity? Could it be both, or perhaps an identity in relation to another operation?

A final important concept is that of an *inverse*.

Definition: Suppose that e is the identity in S with respect to $*$. Then $s \in S$ has **inverse** s' with respect to $*$ if and only if $s * s' = s' * s = e$.

This definition applies to individual elements rather than just one special one or the whole set. Also, it defines a *two-sided* inverse. In some structures, it makes sense to separate left and right inverses—what might that mean? The inverse notion applies in obvious ways for familiar cases, subject to care regarding the binary operation. For instance, in $\mathbb{Z}$ under addition, the identity is 0 and the inverse of 2 is -2, because $2 + (-2) = (-2) + 2 = 0$. In $\mathbb{Z}$ under multiplication, the identity is 1 and the inverse of 2 is $\frac{1}{2}$, because $2 \cdot \frac{1}{2} = \frac{1}{2} \cdot 2 = 1$. Do all elements in $\mathbb{Z}$ have inverses under both operations? Are these inverses in $\mathbb{Z}$? What is the inverse of the identity in each case? And what would be the answers for other sets and binary operations?

To conclude this section, note that it is not a coincidence that I have introduced the definitions of closure, associativity, identities and inverses. As you know if you have read Section 1.3, they all combine in the definition of *group*.

2.3 The definition of *group*

Some definitions in Abstract Algebra are short. But some are longer, with multiple parts. The parts are sometimes called *axioms*: people say things like 'to be a group, a set with a binary operation must satisfy four[1] axioms'. To understand this, it might help to know that axioms can function in two complementary but psychologically distinct ways. The first is as *assumptions*, as things that are obviously true so that everyone can agree to use them without justification. For instance, addition on the real numbers is associative, because for every $x, y, z \in \mathbb{R}$, $(x + y) + z = x + (y + z)$. This is one of numerous axioms for the real numbers, a list of which might be studied in Analysis. There, however, the axioms will not take centre stage for long because the entire subject is about the real numbers, at least at first. This means that the axioms, once clarified, might be used without much comment.

Abstract Algebra, in contrast, is about not one structure but many, and about their similarities and differences. This means that axioms in Abstract Algebra function less as assumptions and more as *criteria*. As we encounter various structures, we ask whether or not they satisfy these criteria. The definition of *group* has four criteria: it requires a set and binary operation that satisfy four axioms.

Definition: A **group** is a set G with a binary operation $*$ such that:

Closure $\forall g_1, g_2 \in G, g_1 * g_2 \in G$;

Associativity $\forall g_1, g_2, g_3 \in G, (g_1 * g_2) * g_3 = g_1 * (g_2 * g_3)$;

Identity $\exists e \in G$ such that $\forall g \in G, e * g = g * e = g$;

Inverses $\forall g \in G, \exists g' \in G$ such that $g * g' = g' * g = e$.

If you have read Chapter 1, you might notice that I have now converted some words to symbols, writing '$\forall$' and '$\exists$' for 'for every' and 'there exists'. Whether and when to do this is debated. Some lecturers think that learning new symbols impedes understanding, or that starting sentences with symbols is grammatically undesirable. Others value the symbols'

---

[1] If your course lists three, see Section 5.8.

brevity. I am of the latter view, though I do not claim full consistency—how I write depends on what I want to communicate. Here, I like that the symbols draw attention to the contrasting structures of the identity and inverses axioms: one uses 'there exists ... for all'; the other uses 'for all ... there exists'. If you have studied some undergraduate mathematics, you might know that order in these *quantifiers* is important. The '∃' comes first in the identity axiom because there is a single identity that combines in a certain way with all other elements. The '∀' comes first in the inverses axiom because each element has its own inverse.

The structure of the group definition could be represented as below, and its length can make Abstract Algebra look daunting—people tend to think that long things are difficult. But that is not necessarily true. Each part of a multi-part definition can be fairly straightforward, so there is no need to panic—we can think about one part at a time. And some parts are simpler than others, so in problem solving it can be useful to think about those first.

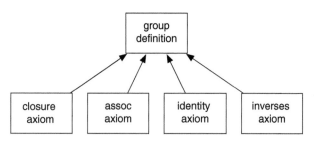

Establishing that a set with a binary operation is a group requires establishing that it satisfies all four group axioms. For instance, the statements below are all true (when the operation is addition, the additive inverse of x is naturally denoted by '$-x$'). So the integers under addition form a group. This group can be denoted by '$(\mathbb{Z}, +)$'.

Closure $\forall x_1, x_2 \in \mathbb{Z}, x_1 + x_2 \in \mathbb{Z}$;

Associativity $\forall x_1, x_2, x_3 \in \mathbb{Z}, (x_1 + x_2) + x_3 = x_1 + (x_2 + x_3)$;

Identity $\exists 0 \in \mathbb{Z}$ such that $\forall x \in \mathbb{Z}, 0 + x = x + 0 = x$;

Inverses $\forall x \in \mathbb{Z}, \exists (-x) \in \mathbb{Z}$ such that $x + (-x) = (-x) + x = 0$.

An Abstract Algebra lecturer might offer similar illustrations and ask students to demonstrate that other structures form groups. But this might leave you feeling nonplussed, at least initially. After all, what do the above statements really do? Just state the obvious, you might argue—they don't really *prove* anything. This psychological conundrum arises from the two complementary functions of axioms, and I will discuss it further in Section 2.4. For now, you might find that the checks make more sense where there is more clearly something to prove.

For instance, the set $3\mathbb{Z} = \{3n \mid n \in \mathbb{Z}\}$ forms a group under addition. Check the axioms to convince yourself: $3\mathbb{Z}$ is closed under addition, addition is associative, the additive identity 0 is in $3\mathbb{Z}$, and every element in $3\mathbb{Z}$ has an additive inverse in $3\mathbb{Z}$. Then think about formulating written arguments. It can help to start by introducing objects, naming them, and specifying their properties. For instance, closure can be established like this.

Claim: $3\mathbb{Z}$ is closed under addition.

Proof: Let $x_1, x_2 \in 3\mathbb{Z}$, so $\exists n_1, n_2 \in \mathbb{Z}$ such that $x_1 = 3n_1, x_2 = 3n_2$.

Then $x_1 + x_2 = 3n_1 + 3n_2 = 3(n_1 + n_2) \in 3\mathbb{Z}$ because $(n_1 + n_2) \in \mathbb{Z}$.

So $3\mathbb{Z}$ is closed under addition.

For the identity axiom, we might write this.

Claim: 0 is the additive identity in $3\mathbb{Z}$.

Proof: $0 = 3 \cdot 0$ so $0 \in 3\mathbb{Z}$, and $\forall x \in 3\mathbb{Z}, 0 + x = x + 0 = x$.

Lecturers' expectations for such arguments vary—some want everything spelled out in full logical detail, others are more relaxed so long as the conceptual ideas come through. If in doubt, show your lecturer some attempts and ask for feedback.

Next, it can be illuminating to prove that a set and binary operation do *not* form a group. Does that require proving that all four axioms are not satisfied? No, it just requires proving that *one* is not satisfied. If that is not obvious, think about the logic. The definition says that to be a group, a structure must satisfy all four axioms. So, to show that it is not a group, it is enough to show that it fails one. This can be quite easy. For instance, the integers do not form a group under multiplication. Scan the axioms— can you find one that is not satisfied? The problem is with inverses. Two elements of $\mathbb{Z}$ do have multiplicative inverses in $\mathbb{Z}$: 1 and -1. But the rest do not, so it is not true that for all $x \in \mathbb{Z}$, there exists $x' \in \mathbb{Z}$ such that $xx' = x'x = 1$.

In cases where some axioms do not hold, it might be possible to restrict to a set on which they do. Consider, for instance, the set of all 2×2 matrices under multiplication. This structure satisfies closure, because multiplying together two 2×2 matrices gives another. Matrix multiplication is associative, which is tedious but straightforward to check. How would you finish this calculation to do so?

$$\left(\begin{pmatrix} a_{11} & a_{12} \\ a_{21} & a_{22} \end{pmatrix} \begin{pmatrix} b_{11} & b_{12} \\ b_{21} & b_{22} \end{pmatrix} \right) \begin{pmatrix} c_{11} & c_{12} \\ c_{21} & c_{22} \end{pmatrix}$$

$$= \begin{pmatrix} a_{11}b_{11} + a_{12}b_{21} & a_{11}b_{12} + a_{12}b_{22} \\ a_{21}b_{11} + a_{22}b_{21} & a_{21}b_{12} + a_{22}b_{22} \end{pmatrix} \begin{pmatrix} c_{11} & c_{12} \\ c_{21} & c_{22} \end{pmatrix}$$

$$= \ldots$$

And 2×2 matrix multiplication has identity $\begin{pmatrix} 1 & 0 \\ 0 & 1 \end{pmatrix}$, because for every matrix $\begin{pmatrix} a & b \\ c & d \end{pmatrix}$,

$$\begin{pmatrix} 1 & 0 \\ 0 & 1 \end{pmatrix} \begin{pmatrix} a & b \\ c & d \end{pmatrix} = \begin{pmatrix} a & b \\ c & d \end{pmatrix} \begin{pmatrix} 1 & 0 \\ 0 & 1 \end{pmatrix} = \begin{pmatrix} a & b \\ c & d \end{pmatrix}.$$

But not all 2×2 matrices have multiplicative inverses. If you have studied matrices, can you list some that do not? Can you suggest a restricted set of matrices that does form a group under multiplication? We will pick up this idea in Section 6.5 and Chapter 9.

2.4 Commutativity and rings

Section 1.2 introduced *commutativity*, which is defined as below.

Definition: The binary operation $*$ is **commutative** on the set S if and only if $\forall s_1, s_2 \in S, s_1 * s_2 = s_2 * s_1$.

Not every binary operation is commutative; matrix multiplication is not, for instance. And commutativity is not part of the definition of *group*. It might seem unnecessary to say that, and I do not suggest that you would forget when asked directly or when stating the definition. But it is easy to forget in general arguments, because everyone is accustomed to switching around the letters in $a + b$ or ab.

Of course, there are plenty of groups in which the operation *is* commutative, and these are are called *abelian* after the mathematician Niels Henrik Abel.

Definition: A group $(G, *)$ is **abelian** if and only if is commutative on G.

Adding a criterion reduces the number of objects that satisfy a definition; there are fewer abelian groups than groups. But many exist—see Chapter 6—and they have some nice, tidy properties. Abelian groups are also pertinent to the definition of *ring*, which appears below.[2] Can you see how?

---

[2] In some areas of mathematics it makes sense to allow rings that do not have multiplicative identities, so you might find that some books or your course omit that axiom.

Definition: A **ring** is a set R with two binary operations $+$ and $\cdot$ such that:

Closure under addition $\forall a, b \in R, a + b \in R$;

Associativity of addition $\forall a, b, c \in R, (a + b) + c = a + (b + c)$;

Additive identity $\exists 0 \in R$ such that $\forall a \in R, 0 + a = a + 0 = 0$;

Additive inverses $\forall a \in R, \exists (-a) \in R$ such that
$a + (-a) = (-a) + a = 0$;

Commutativity of addition $\forall a, b \in R, a + b = b + a$;

Closure under multiplication $\forall a, b \in R, a \cdot b \in R$;

Associativity of multiplication $\forall a, b, c \in R, (a \cdot b) \cdot c = a \cdot (b \cdot c)$;

Multiplicative identity $\exists 1 \in R$ such that $\forall a \in R, 1 \cdot a = a \cdot 1 = a$;

Left distributivity $\forall a, b, c \in R, a \cdot (b + c) = a \cdot b + a \cdot c$;

Right distributivity $\forall a, b, c \in R, (a + b) \cdot c = a \cdot c + b \cdot c$.

This definition really can make Abstract Algebra look nightmarish. Two operations? And who wants to remember that long list of axioms? But it is simpler than it looks because the first five require that R under $+$ be an abelian group. This at least simplifies the theoretical structure.

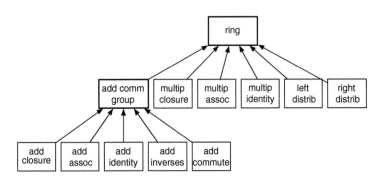

It raises questions, though. Why does addition have to be commutative when multiplication does not? Why do additive inverses have to exist when multiplicative ones do not? That seems a bit arbitrary, no? Of course, mathematicians construct definitions for good reasons, so you should always assume that there is one, even if you do not yet know it. Rings will be discussed in Chapter 9, and it might be useful to know that the canonical example of a ring is $\mathbb{Z}$ with the usual addition and multiplication.

Closure under addition $\forall x, y \in \mathbb{Z}, x + y \in \mathbb{Z}$;

Associativity of addition $\forall x, y, z \in \mathbb{Z}, (x + y) + z = x + (y + z)$;

Additive identity $\exists 0 \in \mathbb{Z}$ such that for every $x \in \mathbb{Z}, 0 + x = x + 0 = 0$;

Additive inverses $\forall x \in \mathbb{Z}, \exists (-x) \in \mathbb{Z}$ such that
$x + (-x) = (-x) + x = 0$;

Commutativity of addition $\forall x, y \in \mathbb{Z}, x + y = y + x$;

Closure under multiplication $\forall x, y \in \mathbb{Z}, x \cdot y \in \mathbb{Z}$;

Associativity of multiplication $\forall x, y, z \in \mathbb{Z}, (x \cdot y) \cdot z = x \cdot (y \cdot z)$;

Multiplicative identity $\exists 1 \in \mathbb{Z}$ such that $\forall a \in \mathbb{Z}, 1 \cdot a = a \cdot 1 = a$;

Left distributivity $\forall x, y, z \in \mathbb{Z}, x \cdot (y + z) = x \cdot y + x \cdot z$;

Right distributivity $\forall x, y, z \in \mathbb{Z}, (x + y) \cdot z = x \cdot z + y \cdot z$.

The psychological issue might now make more sense too. For the integers, these axioms can be understood as assumptions, as things that we all agree are true. We need such assumptions because no one could 'check' that they hold for all integers; we have to take them for granted. They can thus be conceptualized as specifying what the integers *are* by specifying how they behave. If that seems weird, there is no need to worry—you will understand this book if you think of the integers as familiar and the axioms as assumptions that capture their properties. But this accounts for the feeling that establishing that the integers form a group or ring is not really 'proving' anything.

2.5 Mathematical objects and notation

You might have noticed that mathematical objects are subject to notational conventions. Sets are usually denoted by upper-case letters. Groups are often based on sets called G; a second group might be based on a set called H. Subgroups, in particular, might be based on sets called H, or sometimes N or K for *normal subgroups* or the *kernels* of *homomorphisms* (see Chapters 7 and 8). Rings are often based on sets called R.

Elements of sets or groups are usually denoted by lower-case letters. Group elements might be called g and h, or g_1 and g_2, or g and g', or a and b, according to a writer's preferences and what works for a given argument. If an argument involves both a group G and a subgroup H, it might be confusing to use g and h for elements of G; g_1 and g_2 might be better.

Operation notation also varies. A general operation might be denoted by '$*$', but that tends to disappear once Abstract Algebra gets going, for two reasons. First, specific operations often come with notation: '$+$' for addition, '$\times$' or '$\cdot$' or juxtaposition for multiplication, '$\circ$' for function composition. Second, Abstract Algebra involves theory building, constructing general proofs that apply to all groups or all rings. For rings, the two operations are closely linked to standard addition and multiplication so they are usually denoted by '$+$' and either '$\cdot$' or juxtaposition. For groups, there is more variety, which demands notational decisions. To write about all groups, we want notation that does not mislead us into thinking about specific operations; '$*$' is good for that. But mathematicians also value brevity, and dislike writing '$*$' all the time (my handwritten stars always come out wonky, which is annoying). So general statements and arguments tend to be written in multiplicative notation using juxtaposition. With juxtaposition, the group definition looks like this.

Definition: A **group** is a set G with a binary operation such that:

   Closure $\forall g_1, g_2 \in G, g_1 g_2 \in G$;

   Associativity $\forall g_1, g_2, g_3 \in G, (g_1 g_2)g_3 = g_1(g_2 g_3)$;

   Identity $\exists e \in G$ such that $\forall g \in G, eg = ge = g$;

   Inverses $\forall g \in G, \exists g^{-1} \in G$ such that $gg^{-1} = g^{-1}g = e$.

Juxtaposition is economical, and its link to multiplication makes g^{-1} a natural notation for the inverse of g. Using it requires care, though: in an additive group, the operation is still addition, even if this is not explicit. Implicit operations also contribute to economy in speech; people often speak about 'a group G' with no reference to the operation. This is perfectly acceptable, in the same way that using 'if' instead of 'if and only if' in definitions is acceptable. But it can be ill-advised in a novice. If in doubt, specify the operation.

Now, these conventions probably seem familiar, or at least sensible. I highlight them because things do get more complex in Abstract Algebra, and because the subject rewards care in tracking object types. For instance, recall the definition of left coset.

Definition: Suppose that H is a subgroup of G. Then the **left coset** of H containing a is $aH = \{ah | h \in H\}$.

For an additive group, this definition could be written as below. Check that nothing has changed except the operation notation.

Definition: Suppose that H is a subgroup of (an additive group) G. Then the **left coset** of H containing a is $a + H = \{a + h | h \in H\}$.

In these definitions, aH looks like a product and $a + H$ looks like a sum. But they are not products or sums of two *elements*; they are products or sums of *an element and a set*. What kind of object is aH or $a + H$? A coset is defined as a set, but we should be open to the possibility that a set could be empty or contain just one element.

We can grasp the meaning by considering the additive group $G = (\mathbb{Z}, +)$, the element $a = 1$, and the subgroup $H = (3\mathbb{Z}, +)$. This set-up gives

$$a + H = 1 + 3\mathbb{Z} = \{1 + h | h \in 3\mathbb{Z}\} = \{1 + 3n | n \in \mathbb{Z}\}$$
$$= \{\ldots, -5, -2, 1, 4, 7, \ldots\}.$$

What is the left coset containing $a = 2$? How many distinct cosets are there of $(3\mathbb{Z}, +)$ in $(\mathbb{Z}, +)$? How many cosets would there be of the subgroup $(6\mathbb{Z}, +)$ in $(\mathbb{Z}, +)$? Cosets and their important place in theory are discussed in Chapter 7.

To reiterate, I comment on notation not because it is, in itself, difficult or counterintuitive. Notation is always set up sensibly. But, for meaningful understanding, it is important to notice what objects appear in mathematical sentences. Probably the most useful advice anyone[3] gave me about studying Abstract Algebra was that when reading a mathematical sentence, it is a good idea to stop at every object and ask *what kind of object is that?* Reminders to do that appear throughout this book.

---

[3] Thank you, Jean Flower.

CHAPTER 3

Theorems and Proofs

This chapter discusses theorems and proofs in mathematical theories and in undergraduate study. It considers deductions in proofs involving algebra, matrices, modular arithmetic and equivalence relations. It then discusses logic in theorem structures, drawing attention to quantifiers and conditional statements. Its later sections explain how to read proofs effectively, and offer advice on constructing proofs.

3.1 Theorems and proofs in Abstract Algebra

First things first: if you flipped straight to this chapter because you are struggling with theorems and proofs in Abstract Algebra, please start with Chapter 2. People often work on theorems and proofs without fully understanding the underlying axioms and definitions, which makes everything more difficult. If you have read Chapter 2, you will know that I think of mathematical theories as in the diagram below, where the 'bottom' layer contains axioms and definitions. Theorems are proved from these axioms and definitions; they are the 'results' of the deductive science that is mathematics.

But what does it mean to prove theorems or to say that mathematics is deductive? A deductive argument is one in which the *conclusion* is a necessary consequence of the *premises* (also called *assumptions* or *hypotheses*). You might not have thought of mathematics this way, but you are nevertheless accustomed to deductive reasoning. The algebraic step from '$(x - 2)(x - 5) = 0$' to '$x = 2$ or $x = 5$' is a deduction: if the

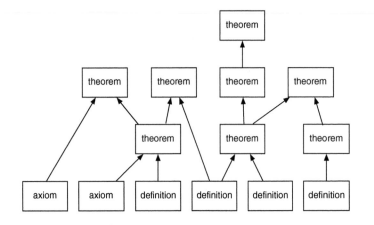

premise that $(x - 2)(x - 5) = 0$ is true, it necessarily follows that $x = 2$ or $x = 5$. A proof chains such deductions together: each step introduces relevant objects or can be justified using axioms, definitions and earlier results. Proofs can be long and complicated, but they can also be short and simple, like this.

Theorem: Suppose that $x^2 - 7x + 10 = 0$. Then $x = 2$ or $x = 5$.

Proof:
$$x^2 - 7x + 10 = 0$$
$$\Rightarrow (x - 2)(x - 5) = 0.$$
$$\text{So } x = 2 \text{ or } x = 5.$$

This means that you have been 'doing' theorems and proofs for some time, even if they haven't been labelled as such. And while specific results like this might be called 'claims' rather than theorems, you have seen general theorems too. For instance, you have used Pythagoras' Theorem, maybe the binomial theorem and, if you have studied complex numbers, maybe De Moivre's Theorem.

However, before undergraduate mathematics—or before upper-level courses in US-like systems—general theorems crop up only occasionally. Students might be expected to engage with the corresponding proofs and perhaps reconstruct them in tests. But they might be expected simply to use the theorems in calculations. Where students do have responsibility for proofs, this often involves proofs of specific types such as proof by induction. Such experience can mean that theorems and proofs seem like part of mathematics, but a smallish and relatively unimportant part. Before university, students expend more effort in a mathematical 'plane' where their primary responsibility is solving problems by doing calculations. They might sometimes be expected to invoke the theorems and proofs that underlie the calculations, but these stay mostly below the level of attention.

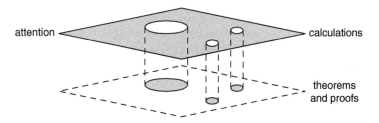

Undergraduate pure mathematics reverses this: calculations become a more minor part of the work, and student attention should be mostly on theorems and proofs.

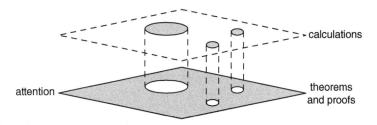

It is important to understand this because probably no one will say it—to mathematicians, it is just obvious that subjects like Abstract Algebra are about theory. And because no one says it, students can be baffled. They expect to be taught problem solving and calculations, and feel let down by lecturers who spend little time on those and much more on theorems and proofs.

In fact, the gap between what students expect and what pure mathematicians do is smaller than it appears: much of the apparent disconnect is pedagogical rather than intellectual. Pure mathematicians do solve problems, but they do it in the theory plane, working out how to use axioms and definitions to prove theorems. And they do not usually start with axioms and definitions and work up. The reality is a messier process of formulating definitions and theorems, tightening up definitions when people use words in slightly different ways, adding conditions when theorems turn out to apply only for some cases, and occasionally reformulating the whole lot when someone notices that a different central concept would permit more elegant proofs. After a while—where a while might be hundreds of years—things settle down enough for everyone to agree about key definitions and important theorems and proofs.

The resulting theory is taught to students. And teaching *does* usually start with axioms and definitions and work up. This makes logical sense, but it means that the psychological experience for students learning a mathematical theory is quite different from that of the mathematicians who developed it. Students do not get a say in formulations: they must accept standard axioms and definitions, and study existing theorems and proofs. This can be uncomfortable, especially for students who want to understand why everything is the way it is—for them, it can seem that definitions and theorems come out of nowhere in a way that is mysterious and therefore annoying. But it is also difficult for dutiful students who are willing to take a lecturer's word—a typical course goes fast, and it takes a lot of work to understand the reasoning packed into theorems and proofs. I say that not to be discouraging—quite the reverse. I want you to be comfortable with the idea that there is much to do.

I also want you to understand that when mathematicians state even short and simple theorems, they have in mind a wealth of related examples and reasoning. Much of Part 2 is designed to provide you with access to something similar: it contains detailed discussions of specific groups and rings, and explains why key concepts are formulated as they are. The present chapter contributes in a different way: it focuses on understanding theorems and proofs as they might be presented in a standard lecture course, with comments at the end on constructing proofs for yourself. Some mathematicians would query that approach—although they teach pre-existing theory, they really want students to master proof construction. But the reality is that most students do a lot of learning from lecture notes and books, and need to know how to interpret these accurately and meaningfully. Accurate and meaningful understanding requires a good grasp of logic, so this chapter proceeds with that.

3.2 Logic in familiar algebra

This section reviews some familiar algebra with a focus on its underlying logic. You might read it and think, *yeah, I know all that*. But many readers will recognize that while they 'know' it in the sense that they could act accordingly, they have not systematically reflected upon this knowledge. This review will build on the discussion in Section 1.2, considering logic in algebra in relation to numbers and matrices.

First, consider equation solving. What would you say it means to solve an equation? Can you get beyond 'finding x'? Maybe think about how you would explain equation solving to a young student who is intelligent but has not yet studied equations. What would you say? I will ask again later.

Perhaps the simplest equations take forms as in Section 1.3: $x + a = b$ or $ax = b$. If these seem trivial, that is due to your extensive knowledge. For young children, the world of numbers is smaller than it is for you, and the equations $x + 5 = 2$ and $5x = 2$ have no solutions. For you, they do have solutions because you know about negative and rational numbers. In Abstract Algebra, we do not revert to the earlier position, but nor do we assume that all numbers are always fair game—we are careful about sets. In the integers, $5x = 2$ has no solution. In the real numbers, $x^2 = -5$ has no solution.

As discussed in Section 1.3, to guarantee that all equations of the form $x + a = b$ can be solved requires that the manipulations below are valid. So it requires a set that is closed under addition and that has associative addition, an additive identity and additive inverses. In short, it requires an additive group.

$$x + a = b$$
$$\Rightarrow (x + a) + (-a) = b + (-a) \quad \text{(adding } -a \text{ to both sides)}$$
$$\Rightarrow x + (a + (-a)) = b + (-a) \quad \text{(reordering operations using associativity)}$$
$$\Rightarrow \qquad x + 0 = b + (-a) \quad \text{(using } a + (-a) = 0\text{)}$$
$$\Rightarrow \qquad x = b + (-a) \quad \text{(using } x + 0 = x\text{)}.$$

To guarantee that all equations of the form $ax = b$ can be solved requires that the manipulations below are valid. So it requires a set that is closed under multiplication and that has associative multiplication, a multiplicative identity and multiplicative inverses. In short, it requires a multiplicative group.

$$ax = b$$
$$\Rightarrow (a^{-1})(ax) = a^{-1}b \quad \text{(multiplying both sides by } a^{-1}\text{)}$$
$$\Rightarrow \quad (a^{-1}a)x = a^{-1}b \quad \text{(reordering operations using associativity)}$$
$$\Rightarrow \qquad 1x = a^{-1}b \quad \text{(using } a^{-1}a = 1\text{)}$$
$$\Rightarrow \qquad x = a^{-1}b \quad \text{(using } 1x = x\text{)}.$$

Now, the equations $x + a = b$ and $ax = b$ each involve just one operation. But algebra often involves addition and multiplication together. The real numbers, for instance, have both operations and many useful properties. They are closed under addition and under multiplication. Both addition and multiplication are associative. There is an additive identity 0, and every element $a \in \mathbb{R}$ has additive inverse $-a \in \mathbb{R}$. There is also a multiplicative identity 1, although multiplicative inverses are not quite so tidy: almost every $a \in \mathbb{R}$ has a multiplicative inverse $a^{-1} \in \mathbb{R}$, but the additive identity 0 does not. Because of this, some simple equations have no solutions: when 0 appears in the multiplicative equation $ax = b$, this equation cannot always be solved for x. The equation $ax = 0$ is mostly fine if uninteresting: provided $a \neq 0$, multiplying both sides by a^{-1} gives $x = 0$. The equation $0x = b$ is mostly not fine: it cannot be solved for x unless $b = 0$. In that case, it reads $0x = 0$, which can be solved but not in the sense of finding 'the' x: the solution set is the whole of $\mathbb{R}$.

Multiplicative equations involving zero can sometimes be solved using not inverses but the *zero product property*. This property was used implicitly in the previous section to deduce that if $(x-2)(x-5) = 0$ then $x-2 = 0$ or $x-5 = 0$. The logic is important, as can be seen by considering this pair of *conditional* statements.

If $ab = 0$ then $a = 0$ or $b = 0$.
If $a = 0$ or $b = 0$ then $ab = 0$.

The first statement is the zero product property: it says that if a product is zero, one of its factors must be zero. The second, its *converse*, is the statement that 'anything times zero is zero'. Do these seem obviously different? The distinction can be hard to keep straight, for two reasons. First, in everyday life, people are sloppy with the word 'if'—they use it in ways that are logically accurate but also in ways that are not.[1] Second, in the real numbers—and thus in most pre-undergraduate equation solving—both statements are true. Such distinctions are easier to keep straight when one statement is true and the other is false. For 2×2 matrices, it is not true that if $ab = 0$ then $a = 0$ or $b = 0$. For instance, neither

$$\begin{pmatrix} 0 & 1 \\ 0 & 0 \end{pmatrix} \text{ nor } \begin{pmatrix} 0 & 2 \\ 0 & 0 \end{pmatrix}$$

is the zero matrix, but

$$\begin{pmatrix} 0 & 1 \\ 0 & 0 \end{pmatrix} \begin{pmatrix} 0 & 2 \\ 0 & 0 \end{pmatrix} = \begin{pmatrix} 0+0 & 0+0 \\ 0+0 & 0+0 \end{pmatrix} = \begin{pmatrix} 0 & 0 \\ 0 & 0 \end{pmatrix}.$$

Formally, matrix multiplication admits *zero divisors*, nonzero elements that multiply to zero. But it is still true that if a or b is the zero matrix, then ab must be the zero matrix.[2]

Another way to clarify the logic of conditional statements is to write in an abbreviated way using symbols (look at the implication arrows).

---

[1] See Section 4.6 of *How to Study for a Mathematics Degree* or its American counterpart *How to Study as a Mathematics Major*.

[2] You might have been told that matrices should always be denoted by upper-case letters. That is a sensible way to distinguish object types, but Abstract Algebra involves reasoning about multiple structures, and no notation matches them all. That said, in specific contexts it is easier to use familiar notation, and we could write 'if $A = 0$ or $B = 0$ then $AB = 0$'.

$$ab = 0 \Rightarrow a = 0 \text{ or } b = 0.$$
$$ab = 0 \Leftarrow a = 0 \text{ or } b = 0.$$

Either way, in logical language, precision matters. Can you see that the three statements below are logically equivalent? For most people, the last requires some thought.

$$ab = 0 \Rightarrow a = 0 \text{ or } b = 0.$$
If $ab = 0$ then $a = 0$ or $b = 0$.
$$ab = 0 \text{ only if } a = 0 \text{ or } b = 0.$$

Precision also matters in equation solving, at least in some contexts. Did you notice that the earlier equations were written $x + a = b$ (x first) but $ax = b$ (x second)? I switched for multiplication because the latter sounds more natural—we usually write $2x$ or $0x$, not $x2$ or $x0$. But the ensuing arguments then involved adding the additive inverse on the right but multiplying by the multiplicative inverse on the left (they are reproduced below so that you can check). Does this matter? Does it make a substantive difference for real numbers or for 2×2 matrices, or is it just convenient to arrive at $(aa^{-1})x$ rather than $(ax)a^{-1}$?

$$x + a = b$$
$$\Rightarrow (x + a) + (-a) = b + (-a) \quad \text{(adding } -a \text{ to both sides)}$$
$$\Rightarrow x + (a + (-a)) = b + (-a) \quad \text{(reordering operations using associativity)}$$
$$\Rightarrow \qquad x + 0 = b + (-a) \quad \text{(using } a + (-a) = 0\text{)}$$
$$\Rightarrow \qquad\qquad x = b + (-a) \quad \text{(using } x + 0 = x\text{)}.$$

$$ax = b$$
$$\Rightarrow (a^{-1})(ax) = a^{-1}b \quad \text{(multiplying both sides by } a^{-1}\text{)}$$
$$\Rightarrow \quad (a^{-1}a)x = a^{-1}b \quad \text{(reordering operations using associativity)}$$
$$\Rightarrow \qquad 1x = a^{-1}b \quad \text{(using } a^{-1}a = 1\text{)}$$
$$\Rightarrow \qquad x = a^{-1}b \quad \text{(using } 1x = x\text{)}.$$

It does matter, due to commutativity. For real numbers and 2×2 matrices, addition is commutative. So, for the additive argument, the order is just convenient: an expression like $(-a) + x + a$ could be tidied up with valid algebraic steps. But for matrices, multiplication is not commutative. So, while associativity means that $(ax)a^{-1} = a(xa^{-1})$, the x and the a^{-1} could not then be swapped to give $a(a^{-1}x)$.

Now, it takes discipline to remember that tempting algebraic steps might not work on the left or right, or for zero, or for various other elements. This can make equation solving seem complicated. In fact, though, Abstract Algebra usually involves minimal equation solving—people tend instead to prove general claims. I am not sure whether this makes students' lives easier or harder. On the one hand, you do not get the disheartening experience of equation-solving mistakes. On the other, you do not get the practice that might make it natural to avoid tempting steps in general proofs. Certainly care is required over algebraic steps in proofs like those appearing later in this chapter.

With that in mind, back to the earlier question: what does it mean to solve an equation? Would you now change your response? I ask not because there is a 'right answer', but to encourage you to reflect on your understanding and improve its accuracy. For instance, when describing equation solving, people often say things like 'change sides, change signs'. But, really, nothing changes sides or signs—each algebraic step does the *same* thing to *both* sides. In terms appropriate to Abstract Algebra, we might say that solving an equation involves making deductions that are valid for the relevant sets and operations in order to reveal the solutions. And what is valid is not universal, which will become increasingly clear as you gain experience with groups and rings. For instance, zero divisors like those seen for matrices also exist in *modular arithmetic*.

3.3 Modular arithmetic

Modular arithmetic is more 'numerical' than matrix arithmetic, but it raises similar issues about algebraic structures. And you already know about modular arithmetic—even if you have never heard of it—because you know about clocks. On much of a clock, arithmetic works as usual. For instance, 3 o'clock plus 5 hours is 8 o'clock; informally, we might write $3 + 5 = 8$. Elsewhere, arithmetic works by 'going over' the 12 and starting again at 1. For instance, 9 o'clock plus 5 hours is 2 o'clock.

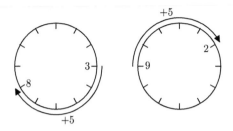

Writing '9 + 5 = 2' would probably feel wrong, though, even if you can see it as 'true' in this structure. So, because this is not 'standard' addition, it has a different name: $+_{12}$ (addition modulo 12). With this notation,

$$3 +_{12} 5 = 8 \text{ and } 9 +_{12} 5 = 2.$$

Now, $+_{12}$ is an operation on a set with just 12 elements, so its effects can be fully captured in a table as below.

$+_{12}$	1	2	3	4	5	6	7	8	9	10	11	12
1	2	3	4	5	6	7	8	9	10	11	12	1
2	3	4	5	6	7	8	9	10	11	12	1	2
3	4	5	6	7	8	9	10	11	12	1	2	3
4	5	6	7	8	9	10	11	12	1	2	3	4
5	6	7	8	9	10	11	12	1	2	3	4	5
6	7	8	9	10	11	12	1	2	3	4	5	6
7	8	9	10	11	12	1	2	3	4	5	6	7
8	9	10	11	12	1	2	3	4	5	6	7	8
9	10	11	12	1	2	3	4	5	6	7	8	9
10	11	12	1	2	3	4	5	6	7	8	9	10
11	12	1	2	3	4	5	6	7	8	9	10	11
12	1	2	3	4	5	6	7	8	9	10	11	12

In this structure, equations can be solved: for instance, $x +_{12} 10 = 4$ has solution $x = 6$ (check against the table and a clock). Indeed, this structure is a group. It can be denoted by $(\mathbb{Z}_{12}, +_{12})$, and relating $(\mathbb{Z}_{12}, +_{12})$ to the additive group of integers $(\mathbb{Z}, +)$ can build intuition for more advanced concepts in Abstract Algebra.

For instance, we need not lose the idea that $9 + 5$ 'really' equals 14. Instead, we can say that 14 is *congruent to* 2 *modulo* 12, writing $14 \equiv 2(\text{mod } 12)$. So the sum

$$9 + 5 = 14 \equiv 2(\text{mod } 12)$$

captures familiar knowledge: 14.00 is 2pm. But, mathematically, we need not be constrained by the clock: it is also true that $26 \equiv 2(\text{mod } 12)$, that $38 \equiv 2(\text{mod } 12)$, and so on. You might find it useful to imagine a bendy integer number line wrapped around a circle so that numbers congruent to one another all line up. Notice that this works for negative numbers too: $-10 \equiv 2(\text{mod } 12)$, $-22 \equiv 2(\text{mod } 12)$ and so on.

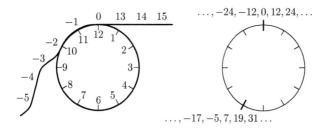

The relation 'congruence modulo 12' thus distributes the integers into disjoint *congruence classes* ('disjoint' means 'not overlapping'). Here are three congruence classes—what are the others?

$$\{\ldots, -24, -12, 0, 12, 24, \ldots\}$$
$$\{\ldots, -23, -11, 1, 13, 25, \ldots\}$$
$$\{\ldots, -22, -10, 2, 14, 26, \ldots\}$$

An integer's congruence class modulo 12 depends on its remainder on division by 12. For instance, $14 \equiv 2(\text{mod } 12)$ because dividing 14 by 12 leaves remainder 2. For this specific case and a general one, we can capture the relationship using the symbols '⇔' ('if and only if' or '(which) is equivalent to') and '∃' ('there exists').

$$14 \equiv 2(\text{mod } 12) \Leftrightarrow \exists n \in \mathbb{Z} \text{ such that } 14 = 2 + 12n;$$
$$x \equiv a(\text{mod } 12) \Leftrightarrow \exists n \in \mathbb{Z} \text{ such that } x = a + 12n.$$

We can thus relate congruence classes to addition modulo 12 via the theorem and proof below. Read both carefully.

Theorem: Suppose that $x \equiv a(\text{mod } 12)$ and $y \equiv b(\text{mod } 12)$.
Then $x + y \equiv (a + b)(\text{mod } 12)$.

Proof: Suppose that $x \equiv a(\text{mod } 12)$ so $\exists n_1 \in \mathbb{Z}$ such that $x = a + 12n_1$
and $\qquad y \equiv b(\text{mod } 12)$ so $\exists n_2 \in \mathbb{Z}$ such that $y = b + 12n_2$.
Then $\qquad x + y = (a + 12n_1) + (b + 12n_2)$
$$= a + (12n_1 + b) + 12n_2 \quad \text{(by associativity)}$$
$$= a + (b + 12n_1) + 12n_2 \quad \text{(by commutativity)}$$
$$= (a + b) + 12(n_1 + n_2) \quad \text{(by associativity)}$$
$$\equiv (a + b)(\text{mod } 12) \qquad \text{(by definition)}.$$

This proof might be longer than some that you have encountered, but its steps are not especially clever. They convert the premises about congruences into information about number forms, then add those number forms and rearrange everything in valid ways to arrive at the conclusion. If you wrote out the theorem and covered up the proof, you could probably reconstruct it. I recommend trying that.

The theorem isn't surprising, either: it says that adding a number congruent to a modulo 12 to one congruent to b modulo 12 always gives one congruent to $(a + b)$ modulo 12. This is algebraically tidy, and familiar from clocks: any representation of 9 o'clock (09.00 or 21.00) plus five hours is always 2 o'clock (14.00 or 02.00). Generalizing, any representation of 9 o'clock plus any number of hours from the congruence class containing 5 (5 plus some number of 12s) is 2 o'clock. Why, exactly? And how does reasoning about a clock relate to the theorem and proof?

Mathematicians capture this theorem's result by saying that addition modulo 12 is *well defined*. This is what makes $9 +_{12} 5 = 2$ meaningful: each 9 in the table for $+_{12}$ can be viewed as representing the congruence class $\{\ldots, -15, -3, 9, 21, 33, \ldots\}$, and each 5 as representing $\{\ldots, -19, -7, 5, 17, 29, \ldots\}$. Adding any element from each always gives an element of $\{\ldots, -22, -10, 2, 14, 26, \ldots\}$. That might seem unexciting, though—when there is pre-existing intuition, it can be hard to see how an operation could fail to be well defined. With less intuition it is often easier, so here it might be useful to consider multiplication. We do not tend to multiply on clocks, but we can. For instance, $4 \times 5 = 20 \equiv 8 \pmod{12}$ so $4 \times_{12} 5 = 8$.

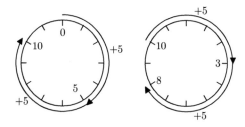

I think it is less obvious that multiplication is well defined. Does multiplying something from the congruence class containing 4 by something from the congruence class containing 5 *always* give something from the congruence class containing 8? Why? Work this out with clocks, then relate that reasoning to the theorem and proof below. Note that the proof uses *distributivity* of multiplication over addition, the property that $x(y + z) = xy + xz$. Where?

Theorem: Suppose that $x \equiv a(\mathrm{mod}\ 12)$ and $y \equiv b(\mathrm{mod}\ 12)$.

Then $xy \equiv (ab)(\mathrm{mod}\ 12)$.

Proof: Suppose that $x \equiv a(\mathrm{mod}\ 12)$ so $\exists n_1 \in \mathbb{Z}$ such that $x = a + 12n_1$

and $\qquad y \equiv b(\mathrm{mod}\ 12)$ so $\exists n_2 \in \mathbb{Z}$ such that $y = b + 12n_2$.

Then $\qquad$
$$
\begin{aligned}
xy &= (a + 12n_1)(b + 12n_2) \\
&= ab + 12n_1 b + 12n_2 a + 12^2 n_1 n_2 \\
&= ab + 12(n_1 b + n_2 a + 12n_1 n_2) \\
&\equiv (ab)(\mathrm{mod}\ 12).
\end{aligned}
$$

This theorem means that $\times_{12}$, like $+_{12}$, is well defined, which is important in ring theory and will be picked up in Chapter 9. But $\times_{12}$, unlike $+_{12}$, does not permit solutions to every simple equation. Consider, for instance,

$$9 \times_{12} x = 3 \qquad 9 \times_{12} x = 1$$

The first equation has solution $x = 3$ because $9 \times 3 = 27 \equiv 3(\mathrm{mod}\ 12)$. Does it have other solutions? The second has no solution because no multiple of 9 is congruent to 1 modulo 12: every multiple of 9 is congruent to 9, 6, 3 or 0. In particular, $9 \times_{12} 4 = 0$, so 9 is a zero divisor. Thus the zero product property—if $ab = 0$ then $a = 0$ or $b = 0$—does not hold for multiplication modulo 12. Are there other zero divisors too? How about for multiplication modulo 8 or modulo 7? Think about that and the zero product property should take its proper place in your mind, not as a general fact but as a theorem that holds for some operations on some sets.

3.4 Equivalence classes

Modular arithmetic forms the basis for an important class of structures in Abstract Algebra. One common notation for these structures focuses not on remainder relationships for individual elements but on congruence classes as sets. For instance, for congruence modulo 12, the congruence class containing 0 is denoted $12\mathbb{Z}$ because $12\mathbb{Z} = \{12n | n \in \mathbb{Z}\} = \{\ldots, -24, -12, 0, 12, 24, \ldots\}$. In fact, $12\mathbb{Z}$ is a *subgroup* of the additive group $\mathbb{Z}$, because it is a subset of $\mathbb{Z}$ that is a group in its own right. Subgroups will be explored in Chapter 6. For now, extending the notation, it makes sense to denote other congruence classes by

$$1 + 12\mathbb{Z} = \{1 + 12n | n \in \mathbb{Z}\} = \{\ldots, -23, -11, 1, 13, 25, \ldots\},$$
$$2 + 12\mathbb{Z} = \{2 + 12n | n \in \mathbb{Z}\} = \{\ldots, -22, -10, 2, 14, 26, \ldots\}, \text{ and so on.}$$

Also, for notational consistency, $12\mathbb{Z}$ can be written

$$0 + 12\mathbb{Z} = \{0 + 12n | n \in \mathbb{Z}\} = \{\ldots, -24, -12, 0, 12, 24, \ldots\}.$$

These congruence classes are *cosets* of the subgroup $12\mathbb{Z}$ in $\mathbb{Z}$, where cosets were introduced briefly in Section 2.6 and will be explored in Chapter 7. Here, coset notation like $1 + 12\mathbb{Z}$ permits a reframing of the previous section because

$$14 \equiv 2 (\mathrm{mod}\ 12) \Leftrightarrow 14 \in 2 + 12\mathbb{Z} \text{ and, in general,}$$
$$x \equiv a (\mathrm{mod}\ 12) \Leftrightarrow x \in a + 12\mathbb{Z}.$$

Thus coset addition and multiplication are well defined: it is meaningful to write
$$(a + 12\mathbb{Z}) + (b + 12\mathbb{Z}) = (a + b) + 12\mathbb{Z} \text{ and}$$
$$(a + 12\mathbb{Z})(b + 12\mathbb{Z}) = (ab) + 12\mathbb{Z}.$$

Now, it might be useful to know that congruence modulo 12 defines an *equivalence relation* on $\mathbb{Z}$ with the cosets as *equivalence classes*. This book could manage without these notions, but many undergraduate degree programmes introduce equivalence relations before Abstract Algebra in a course called something like *Foundations* or *Introduction to Reasoning* or *Sets and Proofs*. So Abstract Algebra lecturers might assume that equivalence relations are familiar and prove theorems about cosets by quoting theorems about equivalence classes. I will not do that, but I will discuss equivalence relations now to highlight some links. (if these notions are unfamiliar and you find the remainder of this section difficult, do skip to the next)

It probably 'sounds right' to speak of congruence in terms of equivalence, saying things like '14 is equivalent to 2 modulo 12'. And this is consistent with theory, because congruence modulo 12 satisfies the criteria below (if these take effort to process, try thinking in terms of remainders on division by 12).

Reflexivity $\forall x \in \mathbb{Z}, x \equiv x (\text{mod } 12)$;
Symmetry $\forall x, y \in \mathbb{Z}$, if $x \equiv y (\text{mod } 12)$ then $y \equiv x (\text{mod } 12)$;
Transitivity $\forall x, y, z \in \mathbb{Z}$, if $x \equiv y (\text{mod } 12)$ and $y \equiv z (\text{mod } 12)$
   then $x \equiv z (\text{mod } 12)$.

This means that congruence modulo 12 is an *equivalence relation* on $\mathbb{Z}$, because it satisfies the definition below. In this definition, '$x \sim y$' can be read as 'x is related to y', where '$\sim$' is a tilde (but I have heard mathematicians call it 'twiddles').

Definition: A relation $\sim$ on a set X is an **equivalence relation** if and only if it is:

Reflexive $\forall x \in X, x \sim x$;
Symmetric $\forall x, y \in X$, if $x \sim y$ then $y \sim x$;
Transitive $\forall x, y, z \in X$, if $x \sim y$ and $y \sim z$ then $x \sim z$.

The notation $x \sim y$ raises potential confusion because it looks a lot like $x * y$ or $x + y$. But a relation on X is not an operation on X. For an operation, every two elements $x, y \in X$ can be combined to give another. For a relation, nothing is combined. Rather, two elements $x, y \in X$ might be related, in which case we write $x \sim y$, or not related, in which case we write $x \not\sim y$. For instance, $14 \equiv 2 (\text{mod } 12)$ but $15 \not\equiv 2 (\text{mod } 12)$.

A simple equivalence relation is $=$ on the set $\mathbb{Q}$: this is reflexive, symmetric and transitive. But not all relations are equivalence relations: the relation $<$ on the set $\mathbb{Q}$ is not reflexive because it is not true that $\forall x \in \mathbb{Q}$, $x < x$ (the relation $<$ is not symmetric either, though it is transitive). Thus equivalence relations are special, and one consequence of their definition is that every equivalence relation *partitions* its set into *equivalence classes*: it distributes the elements into disjoint subsets each containing a full set of equivalent elements. Congruence modulo 12 partitions the integers into congruence classes according to their remainders modulo 12. The relation $=$ on $\mathbb{Q}$ partitions the rational numbers into equivalence classes including $\left\{ \frac{1}{2}, \frac{2}{4}, \frac{3}{6}, \ldots \right\}$, $\left\{ \frac{1}{3}, \frac{2}{6}, \frac{3}{9}, \ldots \right\}$, and so on. In general, we could imagine a partition as below, where the relational statements appear on the right (and pairs not listed are not related—for example, $a_1 \not\sim b_1$). You can check that this arrangement satisfies reflexivity, symmetry and transitivity.

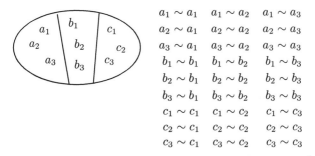

$$
\begin{array}{lll}
a_1 \sim a_1 & a_1 \sim a_2 & a_1 \sim a_3 \\
a_2 \sim a_1 & a_2 \sim a_2 & a_2 \sim a_3 \\
a_3 \sim a_1 & a_3 \sim a_2 & a_3 \sim a_3 \\
b_1 \sim b_1 & b_1 \sim b_2 & b_1 \sim b_3 \\
b_2 \sim b_1 & b_2 \sim b_2 & b_2 \sim b_3 \\
b_3 \sim b_1 & b_3 \sim b_2 & b_3 \sim b_3 \\
c_1 \sim c_1 & c_1 \sim c_2 & c_1 \sim c_3 \\
c_2 \sim c_1 & c_2 \sim c_2 & c_2 \sim c_3 \\
c_3 \sim c_1 & c_3 \sim c_2 & c_3 \sim c_3
\end{array}
$$

The diagram is appropriate for Abstract Algebra because in a finite group, every coset has the same number of elements. But that is neither part nor consequence of the equivalence relation definition—equivalence classes could have different numbers of elements. You will, though, see a proof that every equivalence relation partitions its set. Here I provide a specific version for the theorem below (you can think of the cosets as congruence or equivalence classes).

Theorem: The cosets of $12\mathbb{Z}$ in $\mathbb{Z}$ partition $\mathbb{Z}$.

This is not a theorem for which a proof will make you more convinced. Instead, it should clarify links to wider theory. To understand a proof, I think it helps first to think about the relevant logic. What does it mean to say that cosets partition a set? It means that every element of the set is in a coset, and that the cosets are disjoint. For $12\mathbb{Z}$ in $\mathbb{Z}$, every element of $\mathbb{Z}$ is in a coset because every $a \in \mathbb{Z}$ is in $a + 12\mathbb{Z}$. But how to capture the notion of 'disjoint'? One way is to say that for any two cosets, either the *intersection*—the overlap—is empty, or the sets are the same. Mathematicians denote the intersection of two sets X and Y by $X \cap Y$ ('X intersect Y') and the empty set by $\emptyset$. So we want to prove that for any two cosets $a + 12\mathbb{Z}$ and $b + 12\mathbb{Z}$, either $(a + 12\mathbb{Z}) \cap (b + 12\mathbb{Z}) = \emptyset$ or $a + 12\mathbb{Z} = b + 12\mathbb{Z}$.

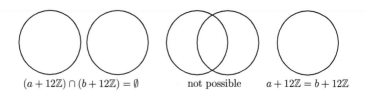

$(a + 12\mathbb{Z}) \cap (b + 12\mathbb{Z}) = \emptyset$ not possible $a + 12\mathbb{Z} = b + 12\mathbb{Z}$

This 'or' statement can be proved by demonstrating that if $(a + 12\mathbb{Z}) \cap (b + 12\mathbb{Z}) \neq \emptyset$, then $a + 12\mathbb{Z} = b + 12\mathbb{Z}$. Think about why, then read the theorem and proof below (note that '$X \subseteq Y$' means 'X is a subset of Y').

Theorem: The cosets of $12\mathbb{Z}$ in $\mathbb{Z}$ partition $\mathbb{Z}$.

Proof: Let $a \in \mathbb{Z}$.

Observe that $0 \in 12\mathbb{Z}$.

So $a = a + 0 \in a + 12\mathbb{Z}$.

Thus every element appears in a coset.

Now suppose that $b \in \mathbb{Z}$ and $(a + 12\mathbb{Z}) \cap (b + 12\mathbb{Z}) \neq \emptyset$.

Then $\exists x \in (a + 12\mathbb{Z}) \cap (b + 12\mathbb{Z})$.

Because $x \in a + 12\mathbb{Z}$, $\exists n_1 \in \mathbb{Z}$ such that $x = a + 12n_1$.

Because $x \in b + 12\mathbb{Z}$, $\exists n_2 \in \mathbb{Z}$ such that $x = b + 12n_2$.

So $a + 12n_1 = b + 12n_2$ and, rearranging,

$b = a + 12(n_1 - n_2)$ and $a = b + 12(n_2 - n_1)$.

Now let $y \in b + 12\mathbb{Z}$.

Then $\exists n_3 \in \mathbb{Z}$ such that $y_1 = b + 12n_3$
$\qquad\qquad\qquad\qquad = a + 12(n_1 - n_2 + n_3) \in a + 12\mathbb{Z}$.

So $b + 12\mathbb{Z} \subseteq a + 12\mathbb{Z}$.

Similarly let $y \in a + 12\mathbb{Z}$.

Then $\exists n_3 \in \mathbb{Z}$ such that $y = a + 12n_3$
$\qquad\qquad\qquad\qquad = b + 12(n_2 - n_1 + n_3) \in b + 12\mathbb{Z}$.

So $a + 12\mathbb{Z} \subseteq b + 12\mathbb{Z}$.

Hence $b + 12\mathbb{Z} \subseteq a + 12\mathbb{Z}$ and $a + 12\mathbb{Z} \subseteq b + 12\mathbb{Z}$.

So $a + 12\mathbb{Z} = b + 12\mathbb{Z}$.

Thus we have proved that

if $(a + 12\mathbb{Z}) \cap (b + 12\mathbb{Z}) \neq \emptyset$ then $a + 12\mathbb{Z} = b + 12\mathbb{Z}$.

Hence the cosets of $12\mathbb{Z}$ in $\mathbb{Z}$ partition $\mathbb{Z}$.

Did you understand that proof to your own satisfaction? If not, or if the length put you off so that you didn't really try, have another go. Take it step by step and, when you get to the end, think about how it all fits together. Guidance on reading proofs effectively is provided in Section 3.6—you might want to reread it after that.

Now, some courses define equivalence relations more formally, introducing an equivalence relation on a set X as a subset of the *Cartesian product* $X \times X$ ('X cross X'). That is pretty abstract, so here is a quick explanation. The set $X \times X$ is the set of all pairs of the form (x, y) where $x, y \in X$. The notation (x, y) might call to mind a plane, which is appropriate because if $X = \mathbb{R}$ then $\mathbb{R} \times \mathbb{R}$ is indeed 'the plane' in the usual sense. If $X = \mathbb{Z}$ then $\mathbb{Z} \times \mathbb{Z}$ is an infinite grid of points with integer coordinates. If X is finite, then $X \times X$ is a finite grid. An equivalence relation is a subset of $X \times X$ because only some elements are related to each other (usually). For instance, for the above equivalence relation on $X = \{a_1, a_2, a_3, b_1, b_2, b_3, c_1, c_2, c_3\}$, the grid below has dots at all points (x, y) for which $x \sim y$. How is reflexivity manifest in the diagram? How about symmetry?

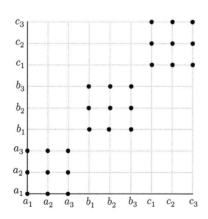

It is not practical to provide a similar diagram for congruence modulo 12 on $\mathbb{Z}$, because $\mathbb{Z} \times \mathbb{Z}$ is an infinite grid and 12 is too big. However, below is a partial version for congruence modulo 3, where the cosets or congruence classes of $3\mathbb{Z}$ in $\mathbb{Z}$ are

$$0 + 3\mathbb{Z} = \{0 + 3x | x \in \mathbb{Z}\} = \{\ldots, -6, -3, 0, 3, 6, 9, \ldots\};$$
$$1 + 3\mathbb{Z} = \{1 + 3x | x \in \mathbb{Z}\} = \{\ldots, -5, -2, 1, 4, 7, 10, \ldots\};$$
$$2 + 3\mathbb{Z} = \{2 + 3x | x \in \mathbb{Z}\} = \{\ldots, -4, -1, 2, 5, 8, 11, \ldots\}.$$

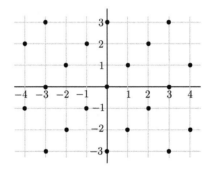

As I say, this way of thinking about equivalence relations is pretty abstract. For the ideas in this book, I find it more useful to represent each element just once and capture relationships in the layout. For instance, for congruence modulo 3, the diagram on the left below is a 'stretched out' version of the bendy number line wrapped around a circle, as if the numbers are spiralling upward so that congruent numbers line up vertically. Each column is a coset; the leftmost is the subgroup $3\mathbb{Z}$. Looking horizontally, each coset is 'offset' from the subgroup by a fixed number. Looking vertically, the elements within every coset differ from one another by elements of the subgroup (in this case, multiples of 3). Chapter 7 will apply this idea to other groups.

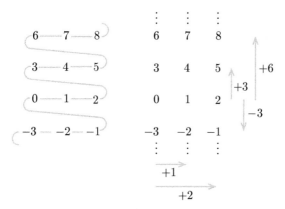

3.5 Logic in theorems

We now return to logic, which underlies all of mathematics and which is important when reading theorems because misinterpreting a theorem leaves you unlikely to understand its proof or to appreciate its place in a theory. First, some specific information on *quantifiers*. When mathematicians talk about quantifiers, they mean

the *universal* quantifier '∀' ('for all'), and
the *existential* quantifier '∃' ('there exists').

The universal quantifier appears in the definition of *associative*.

Definition: The operation ∗ is **associative** on the set S if and only if
$\forall s_1, s_2, s_3 \in S, (s_1 * s_2) * s_3 = s_1 * (s_2 * s_3)$.

This bears on the question in Section 2.3 about whether an operation could be associative on a set but not on a smaller subset. It could not—can you see why? If a property holds for every element of a set, it holds for every element of every subset. Associativity involves three elements, but that makes no difference: if it holds for every three elements of S, it holds for every three elements of every subset of S.

The universal and existential quantifiers appear together in the group identity axiom. As noted in Section 2.3, quantifier order matters: this axiom is about the *existence* of a single element that interacts in a certain way with *all* group elements.

Identity $\exists e \in G$ such that $\forall g \in G$, $e * g = g * e = g$.

To prove that something exists, it is enough to produce one. For instance, in the set of 2×2 matrices, a multiplicative identity is

$$\begin{pmatrix} 1 & 0 \\ 0 & 1 \end{pmatrix}, \text{ because for every matrix } \begin{pmatrix} a & b \\ c & d \end{pmatrix},$$

$$\begin{pmatrix} 1 & 0 \\ 0 & 1 \end{pmatrix}\begin{pmatrix} a & b \\ c & d \end{pmatrix} = \begin{pmatrix} a & b \\ c & d \end{pmatrix}\begin{pmatrix} 1 & 0 \\ 0 & 1 \end{pmatrix} = \begin{pmatrix} a & b \\ c & d \end{pmatrix}.$$

Students sometimes worry about such assertions, though. They feel that it is 'cheating' just to produce an object, that they ought somehow to derive it. But that is not necessary. Justifications are often expected, but it is usually fine effectively to say 'Look, here is one that works, let me prove it'.

Once we know that something exists, we can prove things about it. This might be straightforward: for instance, the axiom implies that an identity commutes with every element in its set. But it might take more work. For instance, your lecturer might prove that for any set and operation, the identity is *unique*, meaning that there is only one identity. For numbers, this might seem weird—of course there is only one '1'. But remember that identities go with operations—there is only one 1, but there is also a 0. And how sure are you that only one 2×2 matrix is a multiplicative identity? With that in mind, read the following theorem and proof. The proof uses a common strategy for proving that something is unique: assume there are two and prove that they must be the same.

Theorem: The identity element for an operation on a set is unique.

Proof: Suppose that e_1 and e_2 are both identity elements for $*$ on S.

Then $\quad e_1 = e_1 * e_2$ because e_2 is an identity

$\qquad\ \ = e_2 \quad$ because e_1 is an identity.

I have always thought this argument rather cute. But it is so short that it can seem vacuous, so it is useful to relate it to some specific sets and operations. For instance, suppose that $S = \mathbb{R}$, the operation is multiplication, $e_1 = 1$, and e_2 is a fictitious other identity element. Then the algebraic part of the argument reads $1 = 1e_2 = e_2$. Similarly, if S is the set of 2×2 matrices, the operation is matrix multiplication,

$$e_1 = \begin{pmatrix} 1 & 0 \\ 0 & 1 \end{pmatrix} \text{ and } \begin{pmatrix} e_a & e_b \\ e_c & e_d \end{pmatrix} \text{ is another identity,}$$

then the algebraic part reads

$$\begin{pmatrix} 1 & 0 \\ 0 & 1 \end{pmatrix} = \begin{pmatrix} 1 & 0 \\ 0 & 1 \end{pmatrix} \begin{pmatrix} e_a & e_b \\ e_c & e_d \end{pmatrix} = \begin{pmatrix} e_a & e_b \\ e_c & e_d \end{pmatrix}.$$

Now, I observed that the identity matrix commutes with all other matrices. Does this contradict the fact that matrix multiplication is not commutative? No, because quantified statements do not have straightforward 'opposites', so defining *not commutative* requires careful *negation*. Commutativity is defined with a universal quantifier.

Definition: The binary operation $*$ is **commutative** on the set S if and only if $\forall s_1, s_2 \in S$, $s_1 * s_2 = s_2 * s_1$.

A naive 'opposite' might be that $\forall s_1, s_2 \in S$, $s_1 * s_2 \neq s_2 * s_1$. But that is not a valid negation. Commutativity fails if $s_1 * s_2 \neq s_2 * s_1$ for even one pair of elements—a single counterexample refutes a universal statement. In the actual negation below, note the new quantifier and think about how it relates to matrices.

Definition: The binary operation $*$ is **not commutative** on the set S if and only if $\exists s_1, s_2 \in S$ such that $s_1 * s_2 \neq s_2 * s_1$.

As explained in Chapter 2, all definitions have *if and only if* structures. But many theorems are conditional statements that hold only in 'one direction'; their converses might be untrue. Section 3.2 considered this via the two statements below. For multiplication on some sets, both are true; for others, just the second is true.

If $ab = 0$ then $a = 0$ or $b = 0$.
If $a = 0$ or $b = 0$ then $ab = 0$.

For practice, consider these statements too. Is each true for all operations on all sets?

If e is the identity then $e * e = e$.
If $e * e = e$, then e is the identity.

The first is true for any identity—why? The second is often not true, even in simple cases. For instance, in the integers, $(-1)(-1) = 1$, but (-1) is not the multiplicative identity. Conditional statements thus merit careful attention, which is challenging in Abstract Algebra for two reasons. First, unsurprisingly, the statements themselves can be complex. Second, the same theorem can often be written in multiple ways. For instance, the following theorems capture the same information about *cyclic groups*. Cyclic groups are discussed in Section 6.3; here, just try to see the logical equivalence.

Theorem: Suppose that G is a cyclic group. Then every subgroup of G is cyclic.

Theorem: Every subgroup of every cyclic group is cyclic.

Theorem: If G is a cyclic group, then every subgroup of G is cyclic.

The first formulation gives the premise and conclusion in separate sentences, one introducing an object and the other saying what can be deduced about it. The second handles everything with universal quantifiers so that the premise and conclusion are packed in together. The third is clearest with regard to conditionality: its structure is '*if* this premise *then* this conclusion'. However, a purist would say that it is not properly delimited and that it should say this.

Theorem: For every group G, if G is a cyclic group, then every subgroup of G is cyclic.

I tend to be a bit slapdash about delimiting; I think it obvious that this theorem is about all groups and find the longer version is a bit of a mouthful. But lack of clear delimiting can introduce ambiguity—does 'if $y > 0$ then …' refer to integers or real numbers, for instance? Certainly it is mathematically polite to introduce everything properly, so attend to your lecturer's approach and maybe err on the side of caution.

Regardless of its formulation, this theorem is true, so the cyclic groups form a subset of groups for which every subgroup is cyclic. But it leaves open the question of whether the converse is also true: if every subgroup of a group is cyclic, must the group be cyclic? Or are the smaller nested sets below in fact the same? If you have studied groups, you might know the answer. If not, you will find it in Chapter 6.

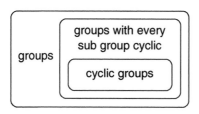

To conclude this section, an important theorem for which there is more mathematical complexity. *Lagrange's Theorem* links the *orders* of finite groups and their *subgroups*, where a subgroup is a subset of a group that is a group in its own right. *Order* means 'size' as in number of elements, and 'divides' is a tidy way to say 'is a factor of'. (If Lagrange's Theorem looks different in your course, see Section 7.7.)

Lagrange's Theorem: Suppose that G is a finite group and H is a subgroup of G. Then the order of H divides the order of G.

The premise of Lagrange's Theorem states that G is a finite group—something like a modular arithmetic group as in Section 3.3—and H is a subgroup. The conclusion says that the order of H divides the order of G. So, for instance, a group of order 12 can have subgroups of order 4 but not of order 5. Does that mean that subsets of order 4 necessarily form subgroups? No: a subset might not contain the identity, for instance.

So the converse of Lagrange's Theorem is not true. The *contrapositive*, however, *is* true, and serves to restrict possible sizes of subgroups within groups.

Statement:
If H is a subgroup of G then the order of H divides the order of G.

Contrapositive:
If the order of H does not divide the order of G, then H is not a subgroup of G.

This is a specific case of the general principle that a conditional statement and its contrapositive are always logically equivalent.

Statement:
If A then B.

Contrapositive:
If not B then not A.

I think about this as follows. Assume that the statement 'If A then B' is true. This means that if B is not true, then A cannot be true because if it were, then B would be true (I do not claim that this is elegant, but I prefer it to more austere, formal explanations). It is useful to be aware of this logical equivalence, because one of a conditional statement and its contrapositive might be easier to work with.

3.6 Self-explanation training

Have you fully understood the theorems and proofs so far in this chapter? I ask because research has shown that students do not always read mathematics effectively. This is not surprising: many are not accustomed to reading mathematics at all. In school, and perhaps in some undergraduate classes, they have watched experts demonstrate how to perform calculations, then practised. They have probably taken notes, but those who understand well might rarely consult those notes, needing only the occasional reminder.

In Abstract Algebra, occasional reading is unlikely to suffice. Most students do not understand entire lectures—the material is too complex and the delivery too fast. So they have to study their notes after class. This sounds like it should be easy because undergraduates have been reading for perhaps 15 years. But mathematical reading is not like ordinary reading. As discussed in the preceding sections, it requires abnormal attention to logic. Moreover, when practised by experts, it is not linear. Mathematicians do not start at the beginning of a page and read until the end. Rather, they seek links among all the words and symbols, reading back and forth along lines of text, between lines, and between paragraphs. They do this even with mathematics that they do not find challenging, because they understand the importance of the logical relationships.

It turns out that students learn to read in a manner more like mathematicians after studying the simple *self-explanation training* below.[3] This training is taken directly from a research project,[4] so its tone is less conversational and more instructional than the rest of this book. It will not turn you into an expert mathematical reader overnight, but implementing it whenever you read mathematical text is likely to improve your understanding.

Self-explanation training

The self-explanation strategy has been found to enhance problem solving and comprehension in learners across a wide variety of academic subjects. It can help you to better understand mathematical proofs: in one research study, students who had worked through these materials before reading a proof scored 30% higher than a control group on a subsequent proof comprehension test.

---

[3] The same training appears in *How to Think about Analysis* so you can skip this if you happen to have read it there. But I'd recommend reviewing it.

[4] For an accessible overview of the research, see Alcock, L., Hodds, M., Roy, S. & Inglis, M. (2015). Investigating and improving undergraduate proof comprehension. *Notices of the American Mathematical Society*, 62, 742–752. For detail, see Hodds, M., Alcock, L., & Inglis, M. (2014). Self-explanation training improves proof comprehension. *Journal for Research in Mathematics Education*, 45, 62–101.

How to self-explain

To improve your understanding of a proof, there is a series of techniques you should apply.

After reading each line:

- Try to identify and elaborate the main ideas in the proof.
- Attempt to explain each line in terms of previous ideas. These may be ideas from the information in the proof, ideas from previous theorems/proofs, or ideas from your own prior knowledge of the topic area.
- Consider any questions that arise if new information contradicts your current understanding.

Before proceeding to the next line of the proof you should ask yourself the following:

- Do I understand the ideas used in that line?
- Do I understand why those ideas have been used?
- How do those ideas link to other ideas in the proof, other theorems, or prior knowledge that I may have?
- Does the self-explanation I have generated help to answer the questions that I am asking?

Below you will find an example showing possible self-explanations generated by students when trying to understand a proof (the labels '(L1)' etc. in the proof indicate line numbers). Please read the example carefully in order to understand how to use this strategy in your own learning.

Example self-explanations

Theorem: No odd integer can be expressed as the sum of three even integers.

Proof: (L1) Assume, to the contrary, that there is an odd integer x such that $x = a + b + c$, where a, b and c are even integers.

(L2) Then $a = 2k, b = 2l$ and $c = 2p$, for some integers k, l and p.

(L3) Thus $x = a + b + c = 2k + 2l + 2p = 2(k + l + p)$.

(L4) It follows that x is even; a contradiction.

(L5) Thus no odd integer can be expressed as the sum of three even integers. □

After reading this proof, one reader made the following self-explanations:

- 'This proof uses the technique of proof by contradiction'.[5]
- 'Since a, b and c are even integers, we have to use the definition of an even integer, which is used in L2.'
- 'The proof then replaces a, b and c with their respective definitions in the formula for x.'
- 'The formula for x is then simplified and is shown to satisfy the definition of an even integer also; a contradiction.'
- 'Therefore, no odd integer can be expressed as the sum of three even integers.'

Self-explanations compared with other comments

You must also be aware that the self-explanation strategy is not the same as *monitoring* or *paraphrasing*. These two methods will not help your learning to the same extent as self-explanation.

Paraphrasing

'a, b and c have to be positive or negative, even whole numbers.'

There is no self-explanation in this statement. No additional information is added or linked. The reader merely uses different words to describe

---

[5] Proof by contradiction is discussed along with other types of proof in Chapter 6 of *How to Study for/as a Mathematics Degree/Major*.

what is already represented in the text by the words 'even integers'. You should avoid using such paraphrasing during your own proof comprehension. Paraphrasing will not improve your understanding of the text as much as self-explanation will.

Monitoring

'OK, I understand that $2(k + l + p)$ is an even integer.'

This statement simply shows the reader's thought process. It is not the same as self-explanation, because the student does not relate the sentence to additional information in the text or to prior knowledge. Please concentrate on self-explanation rather than monitoring.

A possible self-explanation of the same sentence would be:

'OK, $2(k + l + p)$ is an even integer because the sum of 3 integers is an integer and 2 times an integer is an even integer.'

In this example the reader identifies and elaborates the main ideas in the text. They use information that has already been presented to understand the logic of the proof. This is the approach you should take after reading every line of a proof in order to improve your understanding of the material.

That is the end of the self-explanation training. The version used in the research study then provided two theorems and proofs for self-explanation practice. Here, I recommend that you flip back to the theorems and proofs in this chapter and practise with those.

3.7 Writing proofs

Reading proofs more effectively will help you to understand them more fully and become better at constructing proofs of your own. But there is no doubt that proof construction is difficult. Just as mathematical reading is not like ordinary reading, mathematical writing is not like ordinary writing. A proof is not like a message to tell a housemate that you have gone to the supermarket. Nor is it like a standard algorithm that you can learn and apply. There are standard proof strategies, and you should look out for those. But they are not *that* standard—to some extent, each proof is different, which can leave students feeling overwhelmed.

However, specific approaches can help. Some people find it useful to think of proof construction as involving two main processes: a formal part and a problem solving part. The formal part involves using the structure of a theorem to write a 'frame' for its proof: writing the premises, leaving a gap, then writing the conclusion. With that done, it is often possible to work forward from the premises and backward from the conclusion by formulating relevant things in terms of definitions or by making standard or obvious deductions. If you let it, a formal approach will shoulder quite a bit of the burden of proving. Indeed, for simple proofs, it might on its own be enough.

For more complex proofs, you also need problem solving to fill in the gap. This requires insight, which can come from reasoning about familiar examples, or from writing down possibly relevant theorems and thinking about whether they usefully apply. A proof will not flow from your pen in a single stream of mathematically correct argument—probably it will involve some false starts and periods of being stuck, and some cleaning up so that the writing makes sense. But writing one requires no magic. To demonstrate what I mean, I will reason through a formal part and a problem-solving part for this theorem.

Theorem: Suppose that G is a group with identity e and that for every $g \in G$, $g^2 = e$. Then G is abelian.

Is this theorem obviously true? It's not to me. I know that 'abelian' means that the group operation is commutative—that for every $a, b \in G$, $ab = ba$. But I cannot immediately see why that would follow from the fact that squaring every element gives the identity, so the premises and conclusion do not seem obviously linked. It does seem plausible that the squaring property would impose some fairly hefty restrictions on the group, but this is not a theorem that I read and think 'Oh yeah, that must be true'. So proving it requires some work.

My preferred approach to that work is to get some intuition by thinking about examples of groups in which the square of every element is the identity. The problem is that none come to mind, even with a bit of effort.

However, I can always start with a formal approach and see how far that takes me. In this case, I can write a 'frame' for the proof like this.

Theorem: Suppose that G is a group with identity e and that for every $g \in G$, $g^2 = e$. Then G is abelian.

Proof: Suppose that G is a group with identity e and that for every $g \in G$, $g^2 = e$.

. . .

So G is abelian.

At this point I have no idea how to go forward from the premises, but I do know what the line before the conclusion should be because I know how 'abelian' is defined.

Theorem: Suppose that G is a group with identity e and that for every $g \in G$, $g^2 = e$. Then G is abelian.

Proof: Suppose that G is a group with identity e and that for every $g \in G$, $g^2 = e$.

. . .

Thus $\forall a, b \in G, ab = ba$.

So G is abelian.

That prompts a bit of housekeeping because a and b have not been introduced. To fix that, I would add a line after the premises.

Theorem: Suppose that G is a group with identity e and that for every $g \in G$, $g^2 = e$. Then G is abelian.

Proof: Suppose that G is a group with identity e and that for every $g \in G$, $g^2 = e$.

Let $a, b \in G$.

. . .

Thus $\forall a, b \in G, ab = ba$.

So G is abelian.

That's the formal part done, so now I have to work out what could go in the gap. Do you think it will be easier to start at the beginning and try to go forward, or start at the end and try to go backward? My money is on the end, because the claim that $\forall a, b \in G$, $ab = ba$ has more to it. Going backward can go wrong if an argument turns out not to be reversible, but I can try it and keep an eye on the logic. Where could $ab = ba$ come from that would fit with the premises? Well, the premise expression $g^2 = e$ has e on one side, so perhaps I can manipulate $ab = ba$ to get e on one side.

$$
\begin{aligned}
ab = ba &\Leftrightarrow & b^{-1}ab = b^{-1}ba \\
&\Leftrightarrow & b^{-1}ab = a \\
&\Leftrightarrow & a^{-1}b^{-1}ab = a^{-1}a \\
&\Leftrightarrow & a^{-1}b^{-1}ab = e.
\end{aligned}
$$

Now there is an e on the right. Does this help? Not immediately, because I now have a rather messy expression involving some inverses. However, thinking about how to clean those up leads me to notice that, in fact, $g^2 = e \Rightarrow gg = e \Rightarrow g = g^{-1}$. So every element is self-inverse and I can write

$$
\begin{aligned}
ab = ba &\Leftrightarrow a^{-1}b^{-1}ab = e \\
&\Leftrightarrow & abab = e.
\end{aligned}
$$

That does help, because $abab = e$ can be written $(ab)(ab) = e$, which is true by the premise because ab is an element of G. However, I have been working backward, so I need to flip the argument upside down in order to reason from the premises to the conclusion. I can also make it more elegant by explaining the self-inverse idea early on. That leads to something like this.

Theorem: Suppose that G is a group with identity e and that for every $g \in G$, $g^2 = e$. Then G is abelian.

Proof: Suppose that G is a group with identity e and that for every $g \in G$, $g^2 = e$.

Then for every $g \in G$, $g = g^{-1}$.

Let $a, b \in G$.

...

$$abab = e \Leftrightarrow a^{-1}abab = a^{-1}$$
$$\Leftrightarrow \quad bab = a^{-1}$$
$$\Leftrightarrow \quad b^{-1}bab = b^{-1}a^{-1}$$
$$\Leftrightarrow \quad ab = b^{-1}a^{-1}$$
$$\Leftrightarrow \quad ab = ba.$$

So $\forall a, b \in G$, $ab = ba$.

So G is abelian.

That works, but the $abab$ seems to come out of nowhere, so I could introduce it more politely.

Theorem: Suppose that G is a group with identity e and that for every $g \in G$, $g^2 = e$. Then G is abelian.

Proof: Suppose that G is a group with identity e and that for every $g \in G$, $g^2 = e$.

Then for every $g \in G$, $g = g^{-1}$.

Let $a, b \in G$.

Then $ab \in G$ by closure.

So $(ab)(ab) = e$ by the premise.

Now $\quad abab = e \Leftrightarrow a^{-1}abab = a^{-1}$
$$\Leftrightarrow \quad bab = a^{-1}$$
$$\Leftrightarrow \quad b^{-1}bab = b^{-1}a^{-1}$$
$$\Leftrightarrow \quad ab = b^{-1}a^{-1}$$
$$\Leftrightarrow \quad ab = ba.$$

So $\forall a, b \in G$, $ab = ba$.

So G is abelian.

And now I am done, at least in substance. Would you clean up the proof any more, or flesh it out? This is partly a style issue so you are allowed to have opinions about it. Maybe you would really spell out every step, or maybe you would write more tersely. I recommend sharing proof attempts with other students to get a sense of what is possible. If you do that then you will probably agree that a reader needs enough detail to follow the argument, but not so much that they get bogged down.

Of course, you should also attend to what your lecturer values. And be prepared for that to change across courses or within a course: some lecturers expect detail early on but allow flexibility later. Do focus on your writing, though, as a separate cleaning-up task if need be. Mathematicians write in sentences, which means that you should be able to read your work aloud from beginning to end in a way that sounds natural in speech. If you cannot do that then you have probably written some symbols without appropriate surrounding words—work out what words you would add if explaining your argument to someone else, and write them in (maybe consider your punctuation, too).

The important thing to note is that in constructing the above proof, I did not do anything magical—I did not have a massive insight due to extensive experience or luck. I just wrote a proof frame using the premise and conclusion, translated what I knew about these into symbolic form, then looked at what I had and tried something. I didn't write it all forwards in one go—most people cannot do that unless a proof is very similar to one they have seen before; they have to get some ideas on paper, then refine. If the idea of starting before you know how a proof will go makes you nervous, I can only advise you to try it. You will probably surprise yourself.

CHAPTER 4

Studying Abstract Algebra

This chapter covers practical issues in undergraduate study, starting with the transition from earlier mathematics. It discusses strategies for effective learning based on research in cognitive psychology, including self-explanation, interleaving, retrieval practice and spaced study.

4.1 Who are you as a student?

Who are you as a student? Probably you are pretty dedicated: happy to learn, and willing to put in the hours. The problem you will face is that it is easy to be like that for the first two weeks of a course like Abstract Algebra, but hard to sustain for more than about four. By week eight, you might want to lie on the floor, moan quietly and wish for someone to make it all easy. I can't make it easy—undergraduate mathematics just isn't easy. But if you find Abstract Algebra difficult, that is not because you are stupid or incapable. It's because it *is* difficult. If this is your first full-on theorems-and-proofs course, it is likely to seem both difficult and alarmingly different from earlier mathematics. This can make students wonder whether they have topped out—whether they cannot cope with mathematics at this level or, more prosaically, whether they just don't like it.

I would encourage you, though, to avoid making either judgement too soon. Many students transitioning to advanced mathematics have to adjust their expectations in two ways, accepting that they will not understand everything and learning to tolerate longer periods of intellectual discomfort. But most do manage that, and reach a point where they are satisfied with what they have learned. Of course, some then decide that pure mathematics is not their thing and that where possible in future they will avoid it. But better to decide from a position of strength, I think; better to know that you *could* do more but choose not to. Others experience not only new understanding but real joy in trading the more routine aspects of earlier work for logical reasoning and theory building.

To reach a positive position with minimal pain, I think it helps to reflect on your study trajectory, on the decisions you have made. For instance, you probably chose to study at the most prestigious accessible institution. A natural consequence of this is that the material you are taught will be only just within your intellectual reach. If you wished, you could switch to an easier degree or major, switch to a lower-prestige institution, or drop out of higher education and take a different route into professional life. Some students choose to do those things, and more power to them— everyone should think about how to use their time. But most students don't. Most, when they reflect, decide that although it might be difficult, they do want to stick with their degree. Reflection and recommitment help, though. If you recognize that you're doing what you're doing by choice, it becomes easier to put up with its downsides and keep your eye on the prize.

To win the day-to-day battle with the downsides, however, requires thoughtfully managing your studies, which really means thoughtfully managing yourself in relation to your studies. That is because a big part of how well people do in undergraduate mathematics is not intelligence or talent but resilience and self-discipline. I have written about time and self management in *How to Study for a Mathematics Degree* (and the American version *How to Study as a Mathematics Major*) so I will not re-hash that here. What I will do is relate study habits to research on effective learning.

4.2 Myths about learning

Research in cognitive psychology has revealed that people commonly believe some myths about learning. These myths are pervasive because they are based on misleading psychological experience—they are not things that other students tell you, they are things that your own brain 'tells' you. They tend to go unchallenged because few people—including lecturers—are educated about how the human mind works and about what that means for learning. Here I will provide a brief overview focused on specific ways to make learning more effective.

First, your mind. Your mind takes in information from your environment, processes it selectively and works out what to do about the parts that it considers important. The information in educational environments comes from stimuli such as lectures, problem sets, discussions with friends and so on. Your mind processes these stimuli and the result is some kind of behaviour—something to do or write or say.

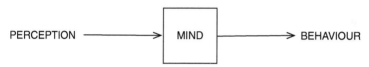

Cognitive psychologists agree that processing takes place in *working memory*, which does the thinking that we experience as thinking.[1] The problem with working memory is that it is small: it can handle only a few pieces of information simultaneously. You will have noticed this if you find, while working on complex problems, that you tend to forget where you started or what you were trying to do.

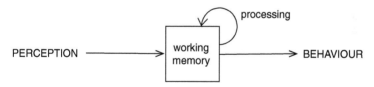

Fortunately, we also have *long-term memory*, which has much, much larger capacity. This is handy, but storing information in long-term memory requires processing in working memory. And using information from long-term memory requires retrieving it.

---

[1] Brains also do a lot of other stuff without involving conscious awareness.

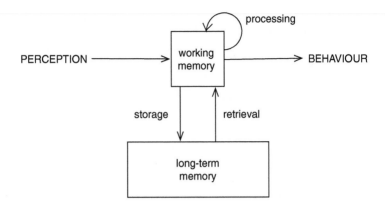

If you have studied psychology then you might already know all of that. But have you considered its implications for learning? Learning involves storing information in long-term memory, and remembering involves retrieving it. How do you think students can make these processes more effective?

Storage is not straightforward because we do not directly control it: wanting to remember something does not guarantee that we will, because learning does not involve simply *recording* stimuli. Rather, it involves *processing* information, interpreting it using existing knowledge; this determines what is stored and how it is interlinked. Similarly, we do not directly control retrieval. We forget things, and muddle them up. This is because memories can be inaccessible due to missing cues—knowledge might be 'in there' but difficult to get out. And they can be muddled due to ambiguous cues—if information is not well differentiated, we might retrieve the wrong things.

You know this, of course. We have all tried and failed to remember names of people we just met, or facts for exams. We have all accidentally used an incorrect word or an inappropriate mathematical procedure. But often things do go right, and we should be encouraged by that. Probably you have learned some mathematics that you could remember immediately because it seemed obvious that it had to work in a certain way. And probably there is mathematics that you do not try to remember because you can reconstruct it whenever you want. If you are like me, this might be part of why you like mathematics: it feels good when things fit together logically and, compared with brute-force memorization, reconstruction is often easier and almost always less boring.

So we all *can* learn new things, indeed very complicated things. And we all *can* store information in a way that enables retrieval. The question is, how can we do that more consistently? I will discuss two strategies for improving storage and two for improving retrieval,[2] relating these to the myths.

First, storage. To store information effectively, we need to process it in what might be called a *deep* way. And often we do not. Many of us have read our notes after a lecture and thought, 'That is definitely my handwriting but I have no recall of it.' So deep processing does not just happen—it often takes effort.

One way to make that effort is to engage in *self-explanation*; mathematical self-explanation training is provided in Section 3.6 (it is short—please read it). Self-explanation might seem unnatural at first because it slows your reading down. But it has been shown to improve learning across a range of academic subjects including proof-based mathematics—it improves both comprehension and retention.[3] Self-explanation works because it leads to more and better connections within a person's knowledge, 'attaching' new information more securely. This naturally also provides more and better differentiated retrieval cues. Self-explanation does not lead to perfect knowledge, but we are not looking for perfection; we are looking for strategies more effective than those people typically use.

A second way to improve storage is to use *interleaving*, switching between topics instead of 'blocking' your learning.

[2] I owe the structure of this section to Iro Xenidou-Dervou, Nina Attridge and Camilla Gilmore, who conduct research in the Centre for Mathematical Cognition at Loughborough University (Nina has since moved to Portsmouth).

[3] See Alcock, L., Hodds, M., Roy, S. & Inglis, M. (2015). Investigating and improving undergraduate proof comprehension. *Notices of the American Mathematical Society*, 62, 742–752, http://www.ams.org/notices/201507/rnoti-p742.pdf.

INTERLEAVING

topic A	topic B	topic C	topic A	topic B	topic C

BLOCKING

topic A	topic A	topic B	topic B	topic C	topic C

Interleaving is almost guaranteed to seem unnatural: most people believe that they learn more by focusing on one thing for an extended time. Moreover, in the short term, *they are right*: blocked study does lead to better immediate performance. But that learning does not tend to stick. Evidence on this comes from research studies like one in which students spent the same amount of time studying formulas for calculating volumes in either interleaved or blocked conditions.[4] During practice, the blockers were considerably more accurate than the interleavers. But during later testing, they had forgotten a large proportion of what they had learned, whereas the interleavers had forgotten almost nothing. It is very important to understand this. Immediate performance is not a good indicator of learning efficacy; good immediate performance does not mean that learning will stick.

In terms of the memory model, interleaving is better because it helps people to notice links between topics and to distinguish similar ideas from different topics. Thus, like self-explanation, it automatically improves retrieval because it leads to more integrated knowledge with more distinguishable cues.

But retrieval can also be enhanced directly. One strategy to use is deliberate *retrieval practice*, which means forcing yourself to recall things. Many students do this already, testing their knowledge before exams, for instance. But most think of it as checking what they know rather than as a way to learn. To learn, they might reread notes or rewatch lectures. But rereading or rewatching is passive, and can give a false sense of security because it involves *recognizing* information, not *recalling* it. Recognizing is much easier—if you have seen something before then it seems familiar,

---

[4] See Rohrer, D. & Taylor, K. (2007). The shuffling of mathematics problems improves learning, *Instructional Science*, 35, 481–498.

which gives you a feeling that you 'know' it. But this can be misleading: it is not the same as recalling information without prompts. Self-testing strengthens retrieval pathways, making future retrieval easier.

A second way to enhance retrieval is to space your study, where spacing is the opposite of cramming or *massing*.

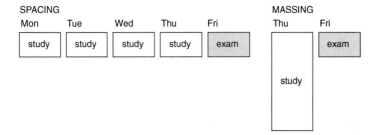

Cramming is common before exams, and some students do it because they are disorganized. But some believe it to be effective. They are wrong, unfortunately. People learn more when they take the same amount of study time and space it out. This is due to the way forgetting works. When we first learn something, we forget it pretty fast. When we review, we forget more slowly. So reviews lead to better retention.

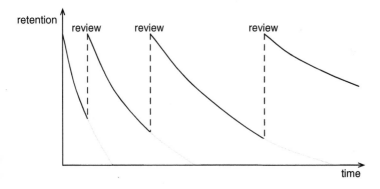

All of this information is linked via an overarching myth that good learning should not feel too hard. People do not necessarily expect learning to be easy, but they do tend to think that smooth, straightforward experience means that learning is going well, and struggle means that it is going badly. This interpretation is generally incorrect. Self-explanation, interleaving, retrieval practice and spacing are likely to feel challenging. But evidence shows that they work.

4.3 Effective learning

The preceding information means that effective learning is *hard*. It is hard to self-explain, to think deeply about links between new material and existing knowledge. It is hard to interleave topics, switching regularly back to things that you have partially forgotten. It is hard to test whether you really can retrieve things, and thus to face the limits of what you know. And it is hard to organize and discipline yourself to space your study across days, weeks and terms or semesters.

So the information in this chapter is not very palatable and I do not expect you to like it. It might be particularly unpalatable if you are one of those people who, until recently, did well in mathematics without making much effort. If that describes you, then probably your ego is tied up with being that person, and adjusting to harder work will be painful. Maybe you will decide not to bother—to disengage from the challenge. Would I respect you less for that? Not really—it is perfectly sensible to decide how hard you are prepared to work for things you want. But I would not want you to disengage for the wrong reasons. Specifically, I would not want you to disengage because you think that people who are really good at mathematics do not have to work hard, do not have to do these things. In fact, they do them routinely.

Professional mathematicians self-explain all the time. Indeed, they consider self-explanation so natural that some are surprised by the idea that students might need training to do it. To them, reading with self-explanation is just proper reading. Of course, they might not always need to know every detail—sometimes it makes sense to trust an expert colleague. But, when learning new mathematics, they self-explain all the time, looking for links to their current understanding.

Mathematicians also routinely interleave and space their work. They might sometimes immerse themselves for long periods in single sets of ideas, but that tends to be exceptional. Many lack the opportunity: most teach and do administration, and must fit their mathematical work around that. Even where prolonged focus is possible, work will probably be broken up. When professional mathematicians write about their craft, they routinely describe working hard and then letting ideas percolate while pursuing some other activity—walking, music, engaging in family life. They know that they have to put in the work, but they also expect insights to come when they are *not* sitting at desks.

Similarly, mathematicians routinely engage in retrieval practice. Again, this is to some extent forced. Mathematicians are often assigned to teach topics that they haven't studied for years. So they have to relearn them. If your lecturers seem to have lots of information at their fingertips, that is not because they picked it up effortlessly the first time and never forgot it. It is because they have revisited it repeatedly so that their forgetting curves are pretty flat. But they also do retrieval practice voluntarily. Mathematicians who have forgotten things will usually try to reconstruct them before looking them up. They instinctively understand the value of retrieving information by rebuilding it from other knowledge.

So what should you do if you decide to suck it up, get over the emotional challenges and study effectively? This is worth some thought because the strategies operate at different levels.

First, you really have to study consistently. Otherwise you will find yourself constantly doing whatever seems most urgent—it will feel impossible to space or interleave your study because today it is vital that you prepare for the next test (or whatever). So a good, realistic routine is beneficial—think about scheduling time for each subject each week, in relatively small blocks so that you revisit everything regularly.

Second, process things deeply. Mathematical study forces this to some extent: most students have to think deeply when solving problems. But you should also do it via self-explanation when reading your notes in order to understand things that you did not grasp in lectures.

Third, test yourself regularly. And note that self-testing can fit well into spaced and interleaved study because it need not be a big operation. In lectures, I sometimes ask students to study a short theorem and proof for two minutes, then cover it up and spend three minutes reconstructing it, then check anything that they missed or found difficult. That provides retrieval practice for a nontrivial course item in six minutes. It probably won't be remembered perfectly for an exam a month later, but it decreases the revision required at that stage.

None of this will do magic, of course. Good strategies will not make mathematics or its study seem easy. But evidence shows that they work.

CHAPTER 5

Binary Operations

This chapter discusses binary operations on numbers and on other objects. It represents the effects of binary operations, contrasting the properties of associativity and commutativity. It then discusses binary operations in relation to modular arithmetic, functions, matrices, symmetries and permutations, observing phenomena to be explored in later chapters. Finally, it discusses binary operations as functions, and technicalities related to closure.

5.1 What is a binary operation?

I f you have read Part 1, you will know that a binary operation takes two elements from a set and combines them to give another. For instance, multiplication on the integers takes two integers and gives another, as in $2 \times 5 = 10$. To capture the effects of a binary operation, tables can be useful. A table cannot show 'all' the results for an infinite set like the integers, but it can visually represent an operation's properties. For instance, in the partial table below we can 'see' that multiplication on the integers is *commutative*, meaning that for every $x, y \in \mathbb{Z}$, $xy = yx$ (see Section 1.2). This is manifest in symmetry about the main diagonal from top left to bottom right. Check that you believe this would hold for the 'whole' table, relating the visual representation to the equation $xy = yx$.

×	0	1	2	3	4	5	6	⋯
0	0	0	0	0	0	0	0	
1	0	1	2	3	4	5	6	
2	0	2	4	6	8	10	12	
3	0	3	6	9	12	15	18	
4	0	4	8	12	16	20	24	
5	0	5	10	15	20	25	30	
6	0	6	12	18	24	30	36	
⋮								

We can also see that 1 is the multiplicative identity for the integers, that for every $x \in \mathbb{Z}$, $1x = x1 = x$ (see Section 2.3). Again, imagine the 'whole' table.

×	0	1	2	3	4	5	6	⋯
0	0	0	0	0	0	0	0	
1	0	1	2	3	4	5	6	
2	0	2	4	6	8	10	12	
3	0	3	6	9	12	15	18	
4	0	4	8	12	16	20	24	
5	0	5	10	15	20	25	30	
6	0	6	12	18	24	30	36	
⋮								

Simple cases can, however, obscure some subtleties. It is easy to assume that the arithmetic operations of addition, multiplication, subtraction and division are unproblematically defined, at least on familiar sets of numbers. But division is not a binary operation on the integers because the expression x/y is not meaningful if $y = 0$; moreover, the set $\mathbb{Z} \backslash \{0\}$ (the integers excluding zero) is not *closed* under division, because it is not true that for every $x, y \in \mathbb{Z}$, $x/y \in \mathbb{Z}$ (see Section 2.2). These properties are captured in the table below, which also highlights the lack of commutativity: for division, it matters which of x and y appears first. Cases of non-commutativity mean that to avoid confusion we need to decide whether the row or column element is x in $x * y$. Which is it, based on x/y below?

÷	0	1	2	3	4	5	6	$\cdots$
0	undefined	0	0	0	0	0	0	
1	undefined	1	$\frac{1}{2}$	$\frac{1}{3}$	$\frac{1}{4}$	$\frac{1}{5}$	$\frac{1}{6}$	
2	undefined	2	1	$\frac{2}{3}$	$\frac{1}{2}$	$\frac{2}{5}$	$\frac{1}{3}$	
3	undefined	3	$\frac{3}{2}$	1	$\frac{3}{4}$	$\frac{3}{5}$	$\frac{1}{2}$	
4	undefined	4	2	$\frac{4}{3}$	1	$\frac{4}{5}$	$\frac{2}{3}$	
5	undefined	5	$\frac{5}{2}$	$\frac{5}{3}$	$\frac{5}{4}$	1	$\frac{5}{6}$	
6	undefined	6	3	2	$\frac{3}{2}$	$\frac{6}{5}$	1	
$\vdots$								

Tables are used extensively in this book because they can both highlight properties and provide intuition for abstract results; this chapter uses them to observe phenomena to be explored formally later. For now, I should say that your course might define a binary operation on a set S as a function on $S \times S$. If that makes sense already, good. If not, it will be discussed along with other formal issues in Section 5.8, after some intuitive ideas.

5.2 Associativity and commutativity

We will start with binary operations on numbers. The standard arithmetic operations are addition, multiplication, subtraction and division, but binary operations can be defined in all sorts of ways. What would tables look like for the operations below, on the integers, say? Are the results always defined? Are the integers closed under each operation? Is each one commutative?

$$a * b = \tfrac{1}{2}(a + b)$$
$$a * b = \sqrt{ab}$$
$$a * b = a$$
$$a * b = 1/(ab)$$
$$a * b = \min\{a, b\} \text{ (the minimum of } a \text{ and } b, \text{ e.g. } \min\{2, -1\} = -1)$$
$$a * b = |a - b|$$

Compared with those of arithmetic, these operations might strike you as contrived. In a sense they are, and the extent to which you encounter such examples will depend upon your lecturer: some use them frequently, others not at all. This book will use them just a little, to consider relationships among binary operations' properties. For instance, consider associativity and commutativity, defined as below. These properties are simple but easy to mix up, or at least to treat ambiguously. Here we will sort out the similarities and differences.

Definition: The operation $*$ is **associative** on the set S if and only if
$$\forall s_1, s_2, s_3 \in S, (s_1 * s_2) * s_3 = s_1 * (s_2 * s_3).$$

Definition: The operation $*$ is **commutative** on the set S if and only if
$$\forall s_1, s_2 \in S, s_1 * s_2 = s_2 * s_1.$$

Associativity and commutativity are similar in that both require an operation and a set (if you have read Part 1 and you are fed up with my remarking on this, you have probably absorbed its importance). Both properties are also about order, but in different ways. Associativity is about the order in which operations are performed, and whether this can be changed without changing the result. Commutativity is about the order in which elements appear, and whether this can be changed without changing the result. This difference is captured in the definitions—check that you see how. But it can be difficult to keep in mind because earlier experience is dominated by arithmetic, where the four standard operations fall neatly into pairs: addition and multiplication are associative and commutative; subtraction and division are neither. For subtraction, for instance,

$$(6 - 3) - 1 \neq 6 - (3 - 1) \quad \text{and} \quad 6 - 2 \neq 2 - 6.$$

Such results can make it seem that associativity and commutativity 'go together', that either we can switch stuff around or we can't. For instance, to calculate $6 + 3 + 7$, we might notice the sum to 10 and think about $6 + (3 + 7)$, implicitly using associativity. If the original sum is written as $3 + 6 + 7$, we might calculate $(3 + 7) + 6$, implicitly using commutativity. These tend to feel like intelligent applications of the same reasoning, so it is useful to recognize that associativity and commutativity do not always go together. For instance, consider the real numbers $\mathbb{R}$ under the binary operation 'take the mean of', which could be denoted by $a * b = \frac{1}{2}(a + b)$. This operation is commutative because

$$\forall a, b \in \mathbb{R}, \quad a * b = \tfrac{1}{2}(a + b) = \tfrac{1}{2}(b + a) = b * a.$$

Is it associative? Is it true that for any three numbers a, b and c, $(a * b) * c = a * (b * c)$? You can probably convince yourself that it is not by looking at the general algebra:

$$(a * b) * c = \tfrac{1}{2}\left(\tfrac{1}{2}(a + b) + c\right);$$
$$a * (b * c) = \tfrac{1}{2}\left(a + \tfrac{1}{2}(b + c)\right).$$

I would also think of number lines.

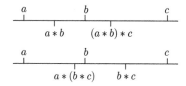

But to really demonstrate, we need a counterexample. For instance,

$$(8 * 4) * 2 = \tfrac{1}{2}\left(\tfrac{1}{2}(8 + 4) + 2\right) = \tfrac{1}{2}(6 + 2) = 4,$$
$$\text{whereas } 8 * (4 * 2) = \tfrac{1}{2}\left(8 + \tfrac{1}{2}(4 + 2)\right) = \tfrac{1}{2}(8 + 3) = \tfrac{11}{2}.$$

Thus 'take the mean of' is commutative but not associative. Is the reverse possible? Can a binary operation be associative but not commutative? To help you think this through, here again is the list of operations. Is each one associative, commutative, both or neither? Check carefully and you will find that all combinations are possible.

$$a * b = \frac{1}{2}(a + b)$$
$$a * b = \sqrt{ab}$$
$$a * b = a$$
$$a * b = 1/(ab)$$
$$a * b = \min\{a, b\}$$
$$a * b = |a - b|$$

Finally, the operation 'take the mean of' can also clarify that binary operations really are *binary*. The operation $*$ defined by $a * b = \frac{1}{2}(a + b)$ takes the mean of *two* numbers, not three, and certainly not more. We could define a different operation to take the mean of three numbers by finding $\frac{1}{3}(a + b + c)$. But that would not be a binary operation, and its results would differ from both $(a * b) * c = \frac{1}{2}\left(\frac{1}{2}(a + b) + c\right)$ and $a * (b * c) = \frac{1}{2}\left(a + \frac{1}{2}(b + c)\right)$. This is important because the expressions $(s_1 * s_2) * s_3$ and $s_1 * (s_2 * s_3)$ involve three elements but apply the binary operation to only two at a time. For associative operations, both orders always give the same result, meaning that $s_1 * s_2 * s_3$ (without brackets) is well defined. If an operation is not associative—or if we do not know whether it is or not—we should take care over which of $(s_1 * s_2) * s_3$ and $s_1 * (s_2 * s_3)$ we write.

5.3 Modular arithmetic

All the sets in this chapter so far have been infinite. But many structures in Abstract Algebra are finite, and an important class of these involves modular arithmetic. You might have studied modular arithmetic in a course on reasoning or number theory, or in Sections 3.3 and 3.4 of this book—please read those now if it is unfamiliar. Then recall that modular arithmetic works like arithmetic on a clock. In arithmetic modulo 12, we say things like '9 plus 5 is 14, which is congruent to 2 modulo 12'. This is written

$$9 + 5 = 14 \equiv 2(\bmod\,12) \quad \text{or} \quad 9 +_{12} 5 = 2.$$

Congruence modulo 12 is based on remainders on division by 12, where

$$14 \equiv 2(\bmod\,12) \iff \exists n \in \mathbb{Z} \text{ such that } 14 = 2 + 12n;$$
$$x \equiv a(\bmod\,12) \iff \exists n \in \mathbb{Z} \text{ such that } x = a + 12n.$$

As proved in Section 3.3, this implies that the operations of addition and multiplication modulo 12 are well defined: if $x \equiv a(\bmod\,12)$ and $y \equiv b(\bmod\,12)$, then

$$x + y \equiv (a + b)(\bmod\,12) \quad \text{and} \quad xy \equiv (ab)(\bmod\,12).$$

As discussed in Section 3.4, congruence modulo 12 is an equivalence relation and therefore *partitions* $\mathbb{Z}$ into disjoint *equivalence classes*. These include

$$12\mathbb{Z} = \quad \{12x | x \in \mathbb{Z}\} = \{\ldots, -24, -12, 0, 12, 24, \ldots\},$$
$$1 + 12\mathbb{Z} = \{1 + 12x | x \in \mathbb{Z}\} = \{\ldots, -23, -11, 1, 13, 25, \ldots\},$$
$$2 + 12\mathbb{Z} = \{2 + 12x | x \in \mathbb{Z}\} = \{\ldots, -22, -10, 2, 14, 26, \ldots\}, \text{ and so on.}$$

In this notation, the addition and multiplication results can be written as

$$(a + 12\mathbb{Z}) + (b + 12\mathbb{Z}) = (a + b) + 12\mathbb{Z};$$
$$(a + 12\mathbb{Z})(b + 12\mathbb{Z}) = (ab) + 12\mathbb{Z}.$$

The *equivalence classes* or *congruence classes* $12\mathbb{Z}, 1 + 12\mathbb{Z}, 2 + 12\mathbb{Z}$ and so on form elements of a set sometimes denoted $\mathbb{Z}_{12}$. Using clock-face numbers to represent these elements, the table below captures the structure $(\mathbb{Z}_{12}, +_{12})$. Every 5 in the table represents the congruence class $\{\ldots, -19, -7, 5, 17, 29, \ldots\}$; this is an instance of compression as described in the Introduction.

$+_{12}$	1	2	3	4	5	6	7	8	9	10	11	12
1	2	3	4	5	6	7	8	9	10	11	12	1
2	3	4	5	6	7	8	9	10	11	12	1	2
3	4	5	6	7	8	9	10	11	12	1	2	3
4	5	6	7	8	9	10	11	12	1	2	3	4
5	6	7	8	9	10	11	12	1	2	3	4	5
6	7	8	9	10	11	12	1	2	3	4	5	6
7	8	9	10	11	12	1	2	3	4	5	6	7
8	9	10	11	12	1	2	3	4	5	6	7	8
9	10	11	12	1	2	3	4	5	6	7	8	9
10	11	12	1	2	3	4	5	6	7	8	9	10
11	12	1	2	3	4	5	6	7	8	9	10	11
12	1	2	3	4	5	6	7	8	9	10	11	12

Notice that for $+_{12}$, the element 12 satisfies the definition of *identity*.

Definition: The element $e \in S$ is the **identity** in S with respect to the binary operation $*$ if and only if $\forall s \in S, e * s = s * e = s$.

Application: The element $12 \in \mathbb{Z}_{12}$ is the identity in $\mathbb{Z}_{12}$ with respect to the binary operation $+_{12}$ because $\forall x \in Z_{12}$, $12 +_{12} x = x +_{12} 12 = x$.

Unfortunately, 12 doesn't 'look like' an identity. But 0 is in the same congruence class and does look like an identity, so it is more commonly used as the representative. The table then appears as below, where 0 falls naturally into the usual identity position in the top row and left column.

$+_{12}$	0	1	2	3	4	5	6	7	8	9	10	11
0	0	1	2	3	4	5	6	7	8	9	10	11
1	1	2	3	4	5	6	7	8	9	10	11	0
2	2	3	4	5	6	7	8	9	10	11	0	1
3	3	4	5	6	7	8	9	10	11	0	1	2
4	4	5	6	7	8	9	10	11	0	1	2	3
5	5	6	7	8	9	10	11	0	1	2	3	4
6	6	7	8	9	10	11	0	1	2	3	4	5
7	7	8	9	10	11	0	1	2	3	4	5	6
8	8	9	10	11	0	1	2	3	4	5	6	7
9	9	10	11	0	1	2	3	4	5	6	7	8
10	10	11	0	1	2	3	4	5	6	7	8	9
11	11	0	1	2	3	4	5	6	7	8	9	10

With an identity in place, we can think about additive inverses.

Definition: Suppose that e is the identity in a set S with respect to the binary operation $*$. Then the element $s \in S$ has inverse s' with respect to $*$ if and only if $s * s' = s' * s = e$.

In a diagram, it is clear that every element has an inverse under $+_{12}$, because there is always an element that 'fills in' the circle.

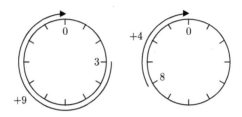

In the table, every element has an inverse because the identity 0 appears in every row and column. In fact, it appears exactly once, and so does every other element. Is that also related to inverses? If so, how? What would go wrong if 0 did not appear in a row, or appeared multiple times?

Next, how about multiplication modulo 12? Using 0 to 11 as representatives, a table for $(\mathbb{Z}_{12}, \times_{12})$ looks like this.

$\times_{12}$	0	1	2	3	4	5	6	7	8	9	10	11
0	0	0	0	0	0	0	0	0	0	0	0	0
1	0	1	2	3	4	5	6	7	8	9	10	11
2	0	2	4	6	8	10	0	2	4	6	8	10
3	0	3	6	9	0	3	6	9	0	3	6	9
4	0	4	8	0	4	8	0	4	8	0	4	8
5	0	5	10	3	8	1	6	11	4	9	2	7
6	0	6	0	6	0	6	0	6	0	6	0	6
7	0	7	2	9	4	11	6	1	8	3	10	5
8	0	8	4	0	8	4	0	8	4	0	8	4
9	0	9	6	3	0	9	6	3	0	9	6	3
10	0	10	8	6	4	2	0	10	8	6	4	2
11	0	11	10	9	8	7	6	5	4	3	2	1

This table is less simple than its counterpart for addition, and it is instructive to compare the two. For instance, 0 is not the identity for $\times_{12}$, because it is not true that $\forall a \in \mathbb{Z}_{12}, \ 0a = a0 = a$ (using juxtaposition to denote $\times_{12}$). Of course, you probably weren't expecting 0 to be the identity for $\times_{12}$; this operation is based on multiplication, so 1 is a more likely candidate. And indeed 1 is the identity because for every $a \in \mathbb{Z}_{12}$, $1a = a1 = a$. However, not every element has an inverse under $\times_{12}$. Some elements do: $5 \times_{12} 5 = 1$, so 5 is *self-inverse*. But 3 has no inverse because there is no element x such that $3x = x3 = 1$. How is that related to the diagram below? Which other elements have multiplicative inverses and which do not? Which elements would have inverses in $(\mathbb{Z}_8, \times_8)$, or in $(\mathbb{Z}_{11}, \times_{11})$?

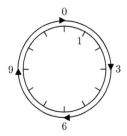

Chapter 9 will pick up this discussion. For now, note that inverses are closely tied to equation solving. You can check from the table that in $(\mathbb{Z}_{12}, \times_{12})$:

the equation $5x = 8$ has exactly one solution $x = 4$;

the equation $3x = 8$ has no solutions;

the equation $3x = 9$ has multiple solutions $x = 3, 7, 11$.

That is because solving these equations relies on multiplicative inverses. To solve an equation like $ax = 8$ for x, we want to multiply both sides (on the left) by a^{-1}. If a^{-1} exists, this is fine. For the first equation,

$$5x = 8 \Rightarrow 5^{-1}5x = 5^{-1}8$$
$$\Rightarrow x = 5^{-1}8$$
$$\Rightarrow x = 5.8 = 4.$$

This gives a unique solution. And, where a^{-1} exists, it will give a unique solution for every equation of the form $ax = b$. So it is not a coincidence that an element has an inverse if and only if every element appears in its row.

If a^{-1} does not exist, then the equation does not have a unique solution; it might have no solutions or multiple solutions, as in $3x = 8$ or $3x = 9$. This differs dramatically from standard multiplication on $\mathbb{R}$, say, where all three equations have unique solutions. So, although you might be accustomed to saying that 'the multiplicative inverse of 5 is $\frac{1}{5}$, that is true only in $\mathbb{Q}$ or $\mathbb{R}$ or $\mathbb{C}$ under standard multiplication. In $(\mathbb{Z}_{12}, \times_{12})$, the inverse of 5 is 5, and some elements do not have inverses. Are there values of n for which every element of the structure $(\mathbb{Z}_n, \times_n)$ has an inverse? No, because 0 never has a multiplicative inverse. Are there values of n for which every element of $(\mathbb{Z}_n \backslash \{0\}, \times_n)$ has an inverse? If so, what are they? Is there an n for which $(\mathbb{Z}_n \backslash \{0\}, \times_n)$ has elements that have inverses but are not self-inverse?

To conclude this section, I would like to make two points. First, it is a while since we touched on subtraction or division. More often in Abstract Algebra we speak about addition or multiplication, then ask whether additive and multiplicative inverses exist. Second, you might have noticed that I have discussed how closure, identities, inverses and commutativity are reflected in binary operation tables. I have not done the same for associativity. Why is that, do you think? There are two reasons. One is that associativity is about triples of elements in expressions like $(s_1 * s_2) * s_3$. A table cannot represent these because it is only two-dimensional—how could a hypothetical three-dimensional 'table' be used to check whether $(s_1 * s_2) * s_3 = s_1 * (s_2 * s_3)$? The other reason is that associativity is typically either axiomatically assumed or *inherited* from a structure in which it is axiomatically assumed. For instance, associativity for addition modulo 12 requires that

$$\forall a, b, c \in \mathbb{Z}, \;\; ((a + b) + c)(\mathrm{mod}\, 12) = (a + (b + c))(\mathrm{mod}\, 12).$$

Why must this be true in terms of remainders and associativity of addition in $\mathbb{Z}$? Checking inherited properties can be tedious, so the extent to which you see it will depend on whether your lecturer is a stickler for detail.

It can require care over quantification, however, and the next section discusses such a case.

5.4 Binary operations on functions

We have so far considered binary operations on numbers and on equivalence classes of numbers. But mathematics also contains binary operations on other objects such as *functions*. Thinking of functions as objects requires more effort than thinking of numbers as objects, because functions are studied later and because a single function can capture lots of information. But language and diagrams help: we talk about 'the function f given by $f(x) = x^2$' ('function' is a noun), and we represent such functions on single graphs.

Like numbers, functions can be combined via a binary operation called addition. This is meaningful only if two functions are defined on the same set, but for plenty of functions that is not problematic. For instance, consider the set of all continuous functions from $\mathbb{R}$ to $\mathbb{R}$, sometimes denoted $C^0(\mathbb{R}, \mathbb{R})$. Two functions $f, g \in C^0(\mathbb{R}, \mathbb{R})$ can be added together:

$$\forall x \in \mathbb{R}, \ (f+g)(x) = f(x) + g(x).$$

Adding two continuous functions always gives another—this might be proved in Calculus or Analysis. So it is true that

$$\forall f, g \in C^0(\mathbb{R}, \mathbb{R}), \ f+g \in C^0(\mathbb{R}, \mathbb{R}).$$

Thus the set $C^0(\mathbb{R}, \mathbb{R})$ is closed under function addition. But notice that this claim uses two layers of quantification. Function addition stipulates that for all $x \in \mathbb{R}$, $(f+g)(x) = f(x) + g(x)$; this means that for all $x \in \mathbb{R}$, the *numbers* $(f+g)(x)$ and $f(x) + g(x)$ are equal. Closure of the set $C^0(\mathbb{R}, \mathbb{R})$ under function addition requires that for all $f, g \in C^0(\mathbb{R}, \mathbb{R})$, $f+g \in C^0(\mathbb{R}, \mathbb{R})$; this involves adding two entire continuous *functions* f and g to get another. For this reason, we tend to tighten up on notation. Before undergraduate level, people often say 'the function $f(x)$', which is ambiguous because $f(x)$ is arguably a number, not a function. I tend to say 'the function f' instead, which helps me to think of f as an element of a set.

Quantification is also important when addressing associativity for function addition on $C^0(\mathbb{R}, \mathbb{R})$. For associativity, we need

$$\forall f, g, h \in C^0(\mathbb{R}, \mathbb{R}), \ \ (f+g) + h = f + (g+h).$$

Is this true? We can check by unpacking the equation into a statement about function equality. The equality $(f+g) + h = f + (g+h)$ means

$$\forall x \in \mathbb{R}, \ \ ((f+g) + h)(x) = (f + (g+h))(x).$$

How would you write a careful argument that this is the case?

Identities and inverses work straightforwardly for addition on $C^0(\mathbb{R}, \mathbb{R})$. However, using '$e$' in the context of functions seems ill advised, so I will switch to the notation 'ι' for this identity. Then $\iota : \mathbb{R} \to \mathbb{R}$ should be a continuous function with the property that

$$\forall f \in C^0(\mathbb{R}, \mathbb{R}), \ \ \iota + f = f + \iota = f.$$

Which function ι has this property? It is $\iota : \mathbb{R} \to \mathbb{R}$ defined by $\iota(x) = 0$ $\forall x \in \mathbb{R}$.

Inverses then follow. The inverse of f is a function $g \in C^0(\mathbb{R}, \mathbb{R})$ with the property that $f + g = g + f = \iota$, meaning that

$$\forall x \in \mathbb{R}, \ \ f(x) + g(x) = g(x) + f(x) = 0.$$

For instance, for $f : \mathbb{R} \to \mathbb{R}$ defined by $f(x) = x^3$, the inverse function is $g : \mathbb{R} \to \mathbb{R}$ defined by $g(x) = -x^3$.

Notice that this is unusual: we normally say that 'the' identity function is $\iota : \mathbb{R} \to \mathbb{R}$ defined by $\iota(x) = x$, and that 'the' inverse of $f : \mathbb{R} \to \mathbb{R}$ defined by $f(x) = x^3$ is $f^{-1} : \mathbb{R} \to \mathbb{R}$ defined by $f^{-1}(x) = \sqrt[3]{x}$. What has happened? As usual, this is about the binary operation. I have been discussing function addition. But when we talk about inverse functions, we usually assume that the binary operation is *composition*, denoted by '$\circ$' or parentheses. Under composition, the identity is $\iota : \mathbb{R} \to \mathbb{R}$ given by $\iota(x) = x$. And $f : \mathbb{R} \to \mathbb{R}$ defined by $f(x) = x^3$ has inverse $f^{-1} : \mathbb{R} \to \mathbb{R}$ defined by $f^{-1}(x) = \sqrt[3]{x}$ because $\forall x \in \mathbb{R}$,

$$f^{-1} \circ f(x) = f^{-1}(f(x)) = \sqrt[3]{x^3} = x$$

and

$$f \circ f^{-1}(x) = f(f^{-1}(x)) = \left(\sqrt[3]{x}\right)^3 = x.$$

Function composition is an important operation for a range of structures in Abstract Algebra, and it is associative. You might see this confirmed as below.

Theorem: Let S be a set and f, g, h be functions mapping S into S.
Then $(f \circ g) \circ h = f \circ (g \circ h)$.

Proof: $\forall x \in S, \ ((f \circ g) \circ h)(x) = (f \circ g)(h(x))$
$$= f(g(h(x)))$$
$$= f(g \circ h(x))$$
$$= (f \circ (g \circ h))(x).$$

5.5 Matrices and transformations

Another type of object is a *matrix*. Matrices, like functions, might feel less like objects than numbers. But matrices of appropriately matching sizes can be combined via binary operations. For instance, for 2×2 matrices, addition and multiplication work like this:

$$\begin{pmatrix} a & b \\ c & d \end{pmatrix} + \begin{pmatrix} k & l \\ m & n \end{pmatrix} = \begin{pmatrix} a+k & b+l \\ c+m & d+n \end{pmatrix};$$

$$\begin{pmatrix} a & b \\ c & d \end{pmatrix} \begin{pmatrix} k & l \\ m & n \end{pmatrix} = \begin{pmatrix} ak+bm & al+bn \\ ck+dm & cl+dm \end{pmatrix}.$$

The set of 2×2 matrices with entries in $\mathbb{R}$ is sometimes denoted by $M_2(\mathbb{R})$ or $M(2, \mathbb{R})$ or $M_{2 \times 2}(\mathbb{R})$ (I know, multiple notations are annoying—I prefer the last as it seems clearest what it means). This set is closed under matrix addition and under matrix multiplication—why? Both operations are associative, which you can check as in Section 1.2. Addition is commutative, but multiplication is not, so $M_{2\times2}(\mathbb{R})$ under multiplication is a non-commutative algebraic structure.

How about identities and inverses? For addition, the identity is a matrix of zeros, and all 2×2 matrices have additive inverses. For multiplication, the identity is

$$\begin{pmatrix} 1 & 0 \\ 0 & 1 \end{pmatrix}, \text{ and } \begin{pmatrix} a & b \\ c & d \end{pmatrix} \text{ has inverse } \frac{1}{ad-bc} \begin{pmatrix} d & -b \\ -c & a \end{pmatrix}.$$

How would you check that these claims are true? And what condition must the *determinant* $ad - bc$ satisfy for a multiplicative inverse to exist?

If you have studied Linear Algebra, you might know that matrices are closely tied to *transformations* or *maps*. A matrix in $M_{2\times2}(\mathbb{R})$ corresponds to a linear map from $\mathbb{R}^2$ to itself. This is because $\mathbb{R}^2$ is an abbreviation for $\mathbb{R} \times \mathbb{R}$, which means the set of all pairs of real numbers (x, y) or the set of all vectors of the form

$$\begin{pmatrix} x \\ y \end{pmatrix} \text{ where } x, y \in \mathbb{R}.$$

Thus $\mathbb{R}^2$ is 'the plane'. A 2×2 matrix A corresponds via matrix multiplication to a linear transformation $T_A : \mathbb{R}^2 \to \mathbb{R}^2$ that maps the plane to itself:

$$\text{if } A = \begin{pmatrix} a_{11} & a_{12} \\ a_{21} & a_{22} \end{pmatrix} \text{ then } A\begin{pmatrix} x \\ y \end{pmatrix} = \begin{pmatrix} a_{11} & a_{12} \\ a_{21} & a_{22} \end{pmatrix}\begin{pmatrix} x \\ y \end{pmatrix} = \begin{pmatrix} a_{11}x + a_{12}y \\ a_{21}x + a_{22}y \end{pmatrix}.$$

Note that the output is an element of $\mathbb{R}^2$ because $a_{11}x + a_{12}y$ and $a_{21}x + a_{22}y$ are both single numbers. Note also that composing transformations corresponds to multiplying matrices: if the matrix B corresponds to another transformation T_B, the matrix for performing T_A then T_B is BA. Why BA rather than AB, and how does that correspond to the usual written order for function composition?

For intuition, I find it helpful to think about which matrices correspond to simple transformations. For instance, the identity matrix maps every point to itself. A multiple of the identity matrix is an enlargement or *dilation* centred at $(0,0)$; this 'makes everything bigger or smaller' because, for instance,

$$\begin{pmatrix} 5 & 0 \\ 0 & 5 \end{pmatrix} \begin{pmatrix} x \\ y \end{pmatrix} = \begin{pmatrix} 5x \\ 5y \end{pmatrix}.$$

What is the determinant for this dilation matrix? What is the inverse transformation and what are its matrix and determinant?

A *projection* to the x-axis sends every point (x, y) to $(x, 0)$; its matrix is shown in the calculation below. What is the corresponding determinant? What geometric problem means that this transformation has no inverse?

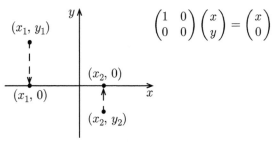

A *reflection* or *flip* in the y-axis retains each point's y-coordinate and sends the x coordinate to the negative of itself. Inspect the diagram and calculation below, then ask what matrix would correspond to a reflection in the x-axis or in the line $y = x$. What are the inverse transformations and what are their matrices and determinants?

A 90° *rotation* about $(0,0)$ takes the point (x, y) to $(-y, x)$. Think about why this works everywhere, not just for (x, y) in the first quadrant.

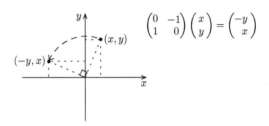

$$\begin{pmatrix} 0 & -1 \\ 1 & 0 \end{pmatrix} \begin{pmatrix} x \\ y \end{pmatrix} = \begin{pmatrix} -y \\ x \end{pmatrix}$$

More generally, a rotation of θ about $(0,0)$ has matrix $\begin{pmatrix} \cos\theta & -\sin\theta \\ \sin\theta & \cos\theta \end{pmatrix}$. What is the inverse of this rotation? What are its matrix and determinant?

If you answered the questions about determinants, you will have observed that all rotations and reflections have determinant ± 1, as do their inverses. Why must a matrix with determinant ± 1 have an inverse with determinant ± 1? Rotations and reflections are *isometries* of the plane, where an isometry is a distance-preserving transformation. *Distance-preserving* means what it sounds like it means: if any two points (x_1, y_1) and (x_2, y_2) are distance d apart before the transformation, they are distance d apart after. Isometries are sometimes called 'rigid motions'—why, do you think? What does this have to do with their determinants?

Finally, you might notice that I have not discussed all possible isometries. As well as rotations and reflections, there exist *translations* and *glide reflections*, which cannot be represented by matrices in the same way and which tend to come up later rather than earlier in Abstract Algebra. Specific rotations and reflections, however, form the elements of an important class of structures, as discussed next.

5.6 Symmetries and permutations

Another type of mathematical object is a *symmetry*. The idea of symmetries as objects was introduced in Section 1.3, which noted that an equilateral triangle has six distinct symmetries: two rotations, three reflections and an identity ('do nothing'). These symmetries are represented below; the dots are just to track which vertex goes where.

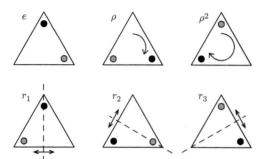

Each symmetry can be understood as a transformation, a function mapping the triangle to itself. We could think of the triangle as centred at $(0, 0)$ so that its symmetries form a subset of isometries of the plane. Indeed, you might observe that the triangle in fact has infinitely many symmetries because we could keep spinning it: a rotation through $480°$ would map it to itself just as well as a rotation through $120°$. But the two have the same effect on the triangle's vertices, so there are only six interestingly different symmetries.

Because symmetries are transformations, they can be combined using composition. For the symmetries of an equilateral triangle, we can construct a binary operation table by cutting out a triangle, labelling its vertices, deciding on a 'start' position, then performing pairs of symmetries to check where it ends up. For instance, as observed in Section 1.3, the rotation ρ followed by the reflection r_1 gives the single reflection r_2.

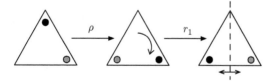

I strongly recommend that you do this cutting, labelling and turning. I am not a stickler for tedious checks, but I do believe that developing intuition for a new structure requires more than just reading a book. It will take five minutes to construct the whole table, so please do. Then check that it matches that below.

$\circ$	e	ρ	ρ^2	r_1	r_2	r_3
e	e	ρ	ρ^2	r_1	r_2	r_3
ρ	ρ	ρ^2	e	r_2	r_3	r_1
ρ^2	ρ^2	e	ρ	r_3	r_1	r_2
r_1	r_1	r_3	r_2	e	ρ^2	ρ
r_2	r_2	r_1	r_3	ρ	e	ρ^2
r_3	r_3	r_2	r_1	ρ^2	ρ	e

Next, take a minute to stare at this table and notice things. The table uses only the six existing symbols, so this set of symmetries is closed under composition; physically, any symmetry puts the triangle back in its 'frame', so performing one then another does too. Second, the table is not symmetrical about its main diagonal, so composition on this set of symmetries is not commutative. There is a subtlety, though, because it is not obvious how to order the elements. I put the rotations 'before' the reflections, but they could be written in any order. Is there an order for which symmetry about the main diagonal would appear? If so, what is it? If not, why not?

As usual, associativity is not visible in the table. You could check it for every triple of elements, but my money would be on your getting bored before you finish. We need not worry, though, because symmetries are transformations and transformations are functions, and function composition is always associative (see Section 5.4). How about identities and inverses? The identity under composition is 'do nothing', which is in the top row and left column as usual. Which elements are inverses of one another? Which are self-inverse? Every element has an inverse, so this set of symmetries under composition forms a group known as the *dihedral group D_3*.

Next, notice that the body of the table splits naturally into four checkerboard squares, two 3×3 squares of rotations (with the identity) and two 3×3 squares of reflections. Why is that, in terms of triangle manipulations?

$\circ$	e	ρ	ρ^2	r_1	r_2	r_3
e	e	ρ	ρ^2	r_1	r_2	r_3
ρ	ρ	ρ^2	e	r_2	r_3	r_1
ρ^2	ρ^2	e	ρ	r_3	r_1	r_2
r_1	r_1	r_3	r_2	e	ρ^2	ρ
r_2	r_2	r_1	r_3	ρ	e	ρ^2
r_3	r_3	r_2	r_1	ρ^2	ρ	e

Mathematically, this pattern manifests a *quotient group*, which will be discussed in Chapter 7. In the meantime, you can develop more intuition by asking how all of this works for the symmetries of a square. How many symmetries does a square have, and how might we denote them? Again, I recommend cutting out and labelling a square and making a binary operation table. Do patterns of inverses work similarly to those for a triangle, and does this table also split naturally into checkerboard squares? Save your table for reference, then decide whether you have enough insight to imagine its equivalent for the symmetries of a regular pentagon or hexagon, or whether it would be worth making tables for those too.

To conclude this section, note that an alternative, number-based labelling provides another way to represent symmetries. Below, for instance, a clockwise rotation ρ of a square through $90°$ sends vertex 1 to position 2, vertex 2 to position 3, and so on; a reflection r_1 in the vertical line swaps vertices 1 and 2 and vertices 3 and 4. Such transformations *permute* the vertices, where a *permutation* of a set S is a rearrangement of the elements of S. Symmetries of the square can be represented as permutations of $\{1, 2, 3, 4\}$.

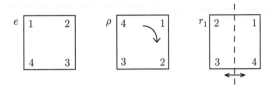

There are two common notations for such permutations. One is 'double-decker' notation, in which the top row of a bracketed array lists the set's elements, and the bottom row shows where each one goes. Permutations corresponding to ρ and r_1 would be denoted

$$\rho = \begin{pmatrix} 1 & 2 & 3 & 4 \\ 2 & 3 & 4 & 1 \end{pmatrix} \quad \text{and} \quad r_1 = \begin{pmatrix} 1 & 2 & 3 & 4 \\ 2 & 1 & 4 & 3 \end{pmatrix}.$$

Mathematicians like brevity, however, and a more economical *tuple* notation uses just one row with one or more sets of brackets. Each element 'goes to' the one that follows in its bracket; the last in any bracket 'goes back around to' the first. So ρ and r_1 would be denoted

$$\rho = (1234) \quad \text{and} \quad r_1 = (12)(34).$$

Formally, a permutation is a *bijection* from a set S to itself—a function whose image is the whole of S, with no two elements mapping to the same element. So two permutations can be treated as objects and combined via composition. The diagram on the right represents this, performing ρ then r_1 for the square. When these are composed, 1 goes to 2 then back to 1 again; 2 goes to 3 then to 4, and so on.

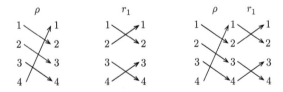

In bracket notations, composition is represented implicitly using juxtaposition. And because permutations are functions, it is common to write the first permutation on the right. We therefore find out 'where an element goes' by following it from the right permutation to the left. Here is how that looks for the element 2 in double-decker notation. How do you think it looks in tuple notation?

$$\begin{pmatrix} 1 & 2 & 3 & 4 \\ 2 & 1 & 4 & 3 \end{pmatrix} \begin{pmatrix} 1 & 2 & 3 & 4 \\ 2 & 3 & 4 & 1 \end{pmatrix} = \begin{pmatrix} 1 & 2 & 3 & 4 \\ 1 & 4 & 3 & 2 \end{pmatrix}$$

Because composing two permutations gives another, the set of all permutations is closed under composition. And, because permutations are functions, composition is associative. What is the identity permutation, and how do inverses work? If shown a permutation, could you write down its inverse?

Finally, consider the relationship between permutations and symmetries. The set of all permutations of $\{1, 2, 3, 4\}$ has $4! = 24$ elements. Why is that? Does every permutation correspond to a symmetry of the square? If not, which do not correspond to symmetries, and why not? We will pick up this discussion in Section 6.9 and Chapter 7.

5.7 Binary operations as functions

This section concludes the chapter by sorting out some formalities. Section 5.1 observed that a binary operation on a set S is formally defined as a function on $S \times S$. What exactly does that mean? It is not about binary operations *on* functions, which take two functions and add or compose them. Rather, it is about binary operations *as* function, on $S \times S$. The notation $S \times S$ means the set of all pairs (s_1, s_2) where $s_1, s_2 \in S$; a function on $S \times S$ assigns an output to every such pair. For instance, a binary operation on $S = \mathbb{R}$ is a function on $\mathbb{R} \times \mathbb{R}$, where $\mathbb{R} \times \mathbb{R}$ is often denoted by $\mathbb{R}^2$. For some readers, functions on $\mathbb{R}^2$ will be familiar from multivariable calculus. For instance, the function $f : \mathbb{R}^2 \to \mathbb{R}$ given by $f(x, y) = x^2 - y^2$ is graphed below: each point (x, y) in the 'input' plane has an 'output' $z = f(x, y)$ at height $x^2 - y^2$.

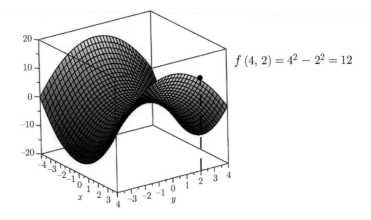

$f(4, 2) = 4^2 - 2^2 = 12$

For a finite-set example, consider $S = \mathbb{Z}_{12}$, as in Section 5.4. A binary operation $*$ on $\mathbb{Z}_{12}$ is a function on $\mathbb{Z}_{12} \times \mathbb{Z}_{12}$, so the 'input' set is a 12×12 grid of points (x, y). Each 'output' is a point above (x, y) at height $x * y$. For instance, the binary operation $\times_{12}$ is a function $\mathbb{Z}_{12} \times \mathbb{Z}_{12} \to \mathbb{Z}_{12}$. It is not practical to graph the 'whole' function, but the diagram below shows points corresponding to

$$3 \times_{12} 1 = 3, \quad 3 \times_{12} 2 = 6, \quad 3 \times_{12} 3 = 9 \quad \text{and} \quad 3 \times_{12} 4 = 0.$$

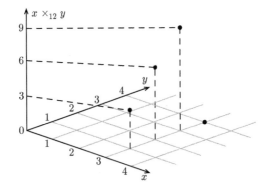

In both cases, a binary operation $*$ on a set S is a function $* : S \times S \to S$. That notation might seem unnatural because we are used to seeing an f

before the colon, not a $*$. So, to highlight the regularity, here are some binary operations considered in this chapter.

$$\times : \mathbb{Z} \times \mathbb{Z} \to \mathbb{Z}$$

$$* : \mathbb{R} \times \mathbb{R} \to \mathbb{R} \text{ given by } a * b = \tfrac{1}{2}(a + b)$$

$$+_{12} : \mathbb{Z}_{12} \times \mathbb{Z}_{12} \to \mathbb{Z}_{12}$$

$$\circ : C^0(\mathbb{R}, \mathbb{R}) \times C^0(\mathbb{R}, \mathbb{R}) \to C^0(\mathbb{R}, \mathbb{R})$$

$$\times : M_{2\times 2}(\mathbb{R}) \times M_{2\times 2}(\mathbb{R}) \to M_{2\times 2}(\mathbb{R})$$

But there subtleties. First, as discussed in Section 5.2, binary operations combine two elements, not three or some other number; that is what it means to describe a binary operation as a function on $S \times S$. In particular, a binary operation is *not* a function on S, so a function like $f : \mathbb{R} \to \mathbb{R}$ given by $f(x) = x^2$ is not a binary operation. That x might be squared, but f has domain $\mathbb{R}$, not $\mathbb{R} \times \mathbb{R}$.

Second, the binary operations listed above are functions $* : S \times S \to S$; their outputs are in the same sets as the elements in their input pairs. This returns us to the point from Section 5.1 about closure: each set is closed under the listed binary operation. And, technically, this has to be the case: for a binary operation $* : S \times S \to S$, the output must be in S. But we have also considered cases in which a set is not closed under a binary operation: $\mathbb{Z} \backslash \{0\}$ is not closed under division, for instance. This does not prevent us talking about division on $\mathbb{Z} \backslash \{0\}$: we could consider division to be a partial function on $\mathbb{Z} \times \mathbb{Z}$ or—as noted in Section 2.2— we could consider division as a binary operation on $\mathbb{Q} \backslash \{0\}$ and note that $\mathbb{Z} \backslash \{0\}$ as a subset of $\mathbb{Q} \backslash \{0\}$ is not closed under division. A binary operation on a restricted subset is sometimes called the *induced* operation on that subset, which is captured in the following, more careful, definition of closure.

Definition: Suppose that $*$ is a binary operation on a set S, and $X \subseteq S$. Then X is **closed** under $*$ if and only if $\forall x_1, x_2 \in X$, $x_1 * x_2 \in X$.

Now, if you digested the information in Section 2.3, you might notice that this formality has implications for the definition of *group*, which is reproduced below (in explicit operation form). Can you see the issue?

Definition: A **group** is a set G with a binary operation $*$ such that:

Closure $\forall g_1, g_2 \in G, g_1 * g_2 \in G$;

Associativity $\forall g_1, g_2, g_3 \in G, (g_1 * g_2) * g_3 = g_1 * (g_2 * g_3)$;

Identity $\exists e \in G$ such that $\forall g \in G, e * g = g * e = g$;

Inverses $\forall g \in G, \exists g' \in G$ such that $g * g' = g' * g = e$.

This definition introduces a set and a binary operation, then lists a closure axiom. But this axiom is redundant, because closure is implicit in the definition of binary operation. Some people, therefore, omit the closure axiom, listing just the other three. But many do list all four, which I think is a good idea. Keeping definitions minimal contributes to mathematical elegance, but it can also lead to overlooking key ideas. Moreover, making closure explicit clarifies links between groups and subgroups, as discussed in the next chapter.

CHAPTER 6

Groups and Subgroups

This chapter introduces groups, including cyclic groups, symmetry groups and permutation groups, as well as groups of numbers and matrices. It discusses group structures via subgroups, generators and relations. It identifies phenomena common across groups and subgroups, for some providing theorems and proofs and for others raising questions to be addressed in later chapters.

6.1 What is a group?

First, a note about language. In everyday life, the words *group* and *set* are used pretty interchangeably. In some situations, one or the other might sound more natural—a group of friends, a set of cutlery—but the meaning is usually a collection of things that go together. In mathematics, this is not the case: the words *group* and *set* have different mathematical meanings. A mathematical *set* is simply a collection of objects. These objects might have something in common, but that is not necessary—a set need not have special properties. A mathematical *group*, on the other hand, does have special properties. A group $(G, *)$ is a set G with a binary operation $*$ that satisfies four[1] axioms as in the definition below (first discussed in Sections 1.3 and 2.3). In Abstract Algebra, we say 'group' only when we mean it in this sense.

---

[1] If the definition in your course has three, see Section 5.7.

Definition: A **group** is a set G with a binary operation $*$ such that:

Closure $\forall g_1, g_2 \in G, g_1 * g_2 \in G$;

Associativity $\forall g_1, g_2, g_3 \in G, (g_1 * g_2) * g_3 = g_1 * (g_2 * g_3)$;

Identity $\exists e \in G$ such that $\forall g \in G, e * g = g * e = g$;

Inverses $\forall g \in G, \exists g' \in G$ such that $g * g' = g' * g = e$.

Section 2.3 established that the integers under the operation of addition satisfy this definition and thus form a group, sometimes denoted $(\mathbb{Z}, +)$. The integer multiples of 3 under addition also form a group, denoted $(3\mathbb{Z}, +)$ where $3\mathbb{Z} = \{3n | n \in \mathbb{Z}\}$. You might want to check that these structures satisfy the group axioms.

Now, although groups involve both sets and binary operations, the sets tend to be more salient. This can mean that students pay insufficient attention to binary operations, which is why this book has a separate binary operations chapter. So, if you flipped straight to the present chapter because you are studying group theory, I recommend that you first flip back and read Chapter 5. Your lecturer might assume that binary operations are simple and familiar, and in a sense they are. But that's what makes them hard to focus on in their own right.

Chapter 5 considered various sets and binary operations in relation to the group properties of closure, associativity, identities and inverses. Do you understand why the structures below all satisfy the group definition? (If you have not read Chapter 5, perhaps this will convince you to do so.)

- the integers under addition
- the integer multiples of 3 under addition
- the integers modulo 12 under addition modulo 12
- the continuous functions on $\mathbb{R}$ under function addition
- the symmetries of an equilateral triangle under composition
- the permutations of the set $\{1, 2, 3, 4\}$ under composition.

Chapter 5 also discussed structures that do *not* form groups. The integers modulo 12 do not form a group under multiplication modulo 12 because not every element has a multiplicative inverse (see Section 5.4). Similarly, the set of all 2×2 matrices does not form a group under multiplication because not every matrix has a multiplicative inverse. Restrictions of these sets that do give groups will be discussed in this chapter.

Before considering specific groups, however, it is worth discussing notation. In the definition above, the binary operation is denoted '$*$' ('star'). But if the operation is (or is 'like') addition, it might be natural to denote the operation by '$+$', the identity by '0' and the inverse of g by '$-g$'. Check that the definition below is equivalent to the general version.

Definition: A **group** is a set G with a binary operation $+$ such that:

> **Closure** $\forall g_1, g_2 \in G, g_1 + g_2 \in G$;
>
> **Associativity** $\forall g_1, g_2, g_3 \in G, (g_1 + g_2) + g_3 = g_1 + (g_2 + g_3)$;
>
> **Identity** $\exists 0 \in G$ such that $\forall g \in G, 0 + g = g + 0 = g$;
>
> **Inverses** $\forall g \in G, \exists (-g) \in G$ such that $g + (-g) = (-g) + g = 0$.

Similarly, as discussed in Section 2.5, the definition can be stated using multiplicative notation, denoting the operation by juxtaposition, the identity by '1' and the inverse of g by 'g^{-1}'. Again, check that the definition below is equivalent to the explicit-operation version. Which version would be most natural for each group listed so far?

Definition: A **group** is a set G with a binary operation (denoted by juxtaposition) such that:

> **Closure** $\forall g_1, g_2 \in G, g_1 g_2 \in G$;
>
> **Associativity** $\forall g_1, g_2, g_3 \in G, (g_1 g_2) g_3 = g_1 (g_2 g_3)$;
>
> **Identity** $\exists 1 \in G$ such that $\forall g \in G, 1g = g1 = g$;
>
> **Inverses** $\forall g \in G, \exists g^{-1} \in G$ such that $g g^{-1} = g^{-1} g = 1$.

Additive and multiplicative notations bring along other conventions. Combining an element with itself can be written using multiples in additive notation, where

$$\underbrace{g + g + \cdots + g}_{n} = ng,$$

and using powers in multiplicative notation, where

$$\underbrace{gg\cdots g}_{n} = g^{n}.$$

In these expressions, g is a group element but n is always an integer, because n versions of the element g are combined. This might seem obvious in groups with non-numerical elements, but it can be harder to track in numerical or general contexts. I will draw attention to possible muddles. Note also that these expressions—without brackets—make sense only because we know that the binary operation in a group is associative (see Section 5.2).

Now, this different-notations business can seem needlessly complicated. A student might wonder why mathematicians don't just pick one notation and stick to it. And in fact they mostly do, but in a considered way. General theorems and proofs tend to be written in multiplicative notation, which is the most economical. But specific notations like '+' for addition or '∘' for composition might be used for specific groups.

And specific groups take up the bulk of this chapter, the main aim of which is to provide familiarity with groups that you will likely encounter. But you should not worry if your course omits some of these or involves different groups—courses have different emphases. A second aim is to draw attention to things worth noticing, in some cases proving that general properties hold for all groups, and in others providing informal introductions to ideas discussed in later chapters. To this end, I will often use group tables (often called *Cayley tables*), which can aid intuition. For instance, the well-trained eye might use them to spot structures such as subgroups.

6.2 What is a subgroup?

What do you think *subgroup* means? If you said 'a subset of a group', that is a good start but not enough. A subgroup is a subset of a group that, with the same binary operation, is a group in its own right. This places restrictions on which subsets are subgroups. For instance, the even integers form a subgroup $(2\mathbb{Z},+)$ of the additive group $(\mathbb{Z},+)$ because they satisfy the four claims below. How would you explain informally what each claim means?

Closure $\forall g_1, g_2 \in 2\mathbb{Z}, g_1 + g_2 \in 2\mathbb{Z}$;

Associativity $\forall g_1, g_2, g_3 \in 2\mathbb{Z}, (g_1 + g_2) + g_3 = g_1 + (g_2 + g_3)$;

Identity $\exists 0 \in 2\mathbb{Z}$ such that $\forall g \in 2\mathbb{Z}, 0 + g = g + 0 = g$;

Inverses $\forall g \in 2\mathbb{Z}, \exists (-g) \in 2\mathbb{Z}$ such that $g + (-g) = (-g) + g = 0$.

In contrast, the odd numbers do not form a subgroup of $(\mathbb{Z},+)$. They do not satisfy closure because adding two odd numbers does not give an odd number. They contain no additive identity because zero is even. The odd numbers do satisfy associativity and the existence of additive inverses, but that is only two of the four axioms. So they do not form a group under addition in their own right and thus cannot form a subgroup of $(\mathbb{Z},+)$.

For a less familiar example, consider D_3, the *dihedral group* formed by the six symmetries of an equilateral triangle under composition. As discussed in Section 5.6, D_3 is a group because it is closed under composition, symmetries are functions so that composition is associative (see Section 5.5), there is an identity element (the 'do nothing' symmetry) and every symmetry has an inverse.

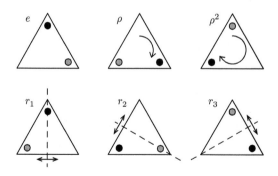

One subgroup of D_3 is $\{e, \rho, \rho^2\}$. The shading in the table below shows this as a self-contained group within D_3. More formally, $\{e, \rho, \rho^2\}$ is closed under composition, composition is associative, and $\{e, \rho, \rho^2\}$ contains the identity and all of its elements' inverses.

$\circ$	e	ρ	ρ^2	r_1	r_2	r_3
e	e	ρ	ρ^2	r_1	r_2	r_3
ρ	ρ	ρ^2	e	r_2	r_3	r_1
ρ^2	ρ^2	e	ρ	r_3	r_1	r_2
r_1	r_1	r_3	r_2	e	ρ^2	ρ
r_2	r_2	r_1	r_3	ρ	e	ρ^2
r_3	r_3	r_2	r_1	ρ^2	ρ	e

As noted in Chapter 5, associativity is no fun to check directly—no one wants to confirm that for every three elements g_1, g_2 and g_3 it is true that $(g_1 * g_2) * g_3 = g_1 * (g_2 * g_3)$. Fortunately, associativity is defined using the single universal quantifier *for all*, so every subset of every group *inherits* associativity from the group (see Section 3.5). This reasoning means that associativity is never a problem for subgroups. And it is often easy to check whether a possible subgroup contains the identity. Inverses and closure might require more thought, though. With that in mind, can you identify other subgroups of D_3? Can you find them all? Try this before moving on as it will help you to get a feel for the structure of this group and for how subgroups work in general. Section 6.7 will examine dihedral groups in more detail, and Section 6.10 will consider conditions guaranteeing that a subset of a group is a subgroup.

In the meantime, you can develop intuition by considering other groups. Below is a table for the group $(\mathbb{Z}_{12}, +_{12})$, the integers modulo 12 under addition modulo 12 (see Section 5.3). The shaded elements form a subgroup: the set $\{0, 3, 6, 9\}$ is closed under $+_{12}$, associativity is inherited, the identity 0 is included, and 3 and 9 are mutually inverse and 0 and 6 are self-inverse. How are these properties manifest in the table?

$+_{12}$	0	1	2	3	4	5	6	7	8	9	10	11
0	0	1	2	3	4	5	6	7	8	9	10	11
1	1	2	3	4	5	6	7	8	9	10	11	0
2	2	3	4	5	6	7	8	9	10	11	0	1
3	3	4	5	6	7	8	9	10	11	0	1	2
4	4	5	6	7	8	9	10	11	0	1	2	3
5	5	6	7	8	9	10	11	0	1	2	3	4
6	6	7	8	9	10	11	0	1	2	3	4	5
7	7	8	9	10	11	0	1	2	3	4	5	6
8	8	9	10	11	0	1	2	3	4	5	6	7
9	9	10	11	0	1	2	3	4	5	6	7	8
10	10	11	0	1	2	3	4	5	6	7	8	9
11	11	0	1	2	3	4	5	6	7	8	9	10

The elements of the subgroup $\{0, 3, 6, 9\}$ are not adjacent in the 'natural' table order, but they can be pulled out to form a mini-table.

$+_{12}$	0	3	6	9
0	0	3	6	9
3	3	6	9	0
6	6	9	0	3
9	9	0	3	6

What other subgroups does $(\mathbb{Z}_{12}, +_{12})$ have? Again, can you list them all? And do you notice anything about the number of elements in each subgroup? Could $(\mathbb{Z}_{12}, +_{12})$ have a subgroup with exactly five elements, for instance? How about eight? Do you have an intuitive sense of why there must be a restriction?

6.3 Cyclic groups and subgroups

The group $(\mathbb{Z}_{12}, +_{12})$ is a *cyclic group*, which might seem a natural description based on the clock-face diagrams in Section 5.3. We can 'cycle through' the whole group by starting at the identity 0 (or, indeed, anywhere else) and repeatedly adding 1.

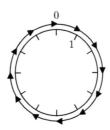

More technically, 1 is a *generator* for this group, because repeatedly combining 1 with itself via the operation $+_{12}$ generates the whole group. Definitions of *generate* and *cyclic group* for additive groups appear below. Notice that different brackets are used to convey specific meanings.

Definition: Let $(G, +)$ be a group and $g \in G$. The set **generated by** g is
$$\langle g \rangle = \{ng | n \in \mathbb{Z}\}.$$

Definition: Let $(G, +)$ be a group. Then G is **cyclic** if and only if $\exists g \in G$ such that $G = \langle g \rangle$.

For groups expressed in multiplicative notation, the first definition appears below. Why does the definition of *cyclic* not need to change?

Definition: Let $(G, \times)$ be a group and $g \in G$. The **set generator** g is
$$\langle g \rangle = \{g^n | n \in \mathbb{Z}\}.$$

Now, does $(\mathbb{Z}_{12}, +_{12})$ have generators other than 1? This question might seem strange because it is intuitively natural to think of 1 as 'the' generator. But there is nothing in the definition to say that a generator has to be unique. Is 2 a generator, for instance?

No, because the set generated by 2 is $\langle 2 \rangle = \{2, 4, 6, 8, 10, 0\} \neq \mathbb{Z}_{12}$. Similarly, neither 3 nor 4 is a generator, because $\langle 3 \rangle = \{3, 6, 9, 0\}$ and $\langle 4 \rangle = \{4, 8, 0\}$. Notice that $\langle 9 \rangle = \langle 3 \rangle$ and $\langle 8 \rangle = \langle 4 \rangle$, and that every element generates a subgroup. Does 5 generate $(\mathbb{Z}_{12}, +_{12})$? Check and you will find that it does.

 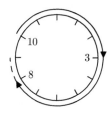

List all generators and you will notice that each is *relatively prime* to 12—the generator and 12 have no common factors. Why?

Next, how do these ideas apply to the cyclic group $(\mathbb{Z}_7, +_7)$, the integers modulo 7 under addition modulo 7? What generators does this have?

$+_7$	0	1	2	3	4	5	6
0	0	1	2	3	4	5	6
1	1	2	3	4	5	6	0
2	2	3	4	5	6	0	1
3	3	4	5	6	0	1	2
4	4	5	6	0	1	2	3
5	5	6	0	1	2	3	4
6	6	0	1	2	3	4	5

Because 7 is prime, every nonzero element is relatively prime to 7 and thus generates the whole group. Does that mean that $(\mathbb{Z}_7, +_7)$ has no subgroups? No: it still has two. The first is the whole group: a group is always a subgroup of itself. If you are inclined to think that the whole group is not a 'proper' subgroup, you will be pleased to learn that mathematicians use exactly that language. A *proper subgroup* is a subgroup that is not the whole group, so $(\mathbb{Z}_7, +_7)$ is not a proper subgroup of itself. The other subgroup of $(\mathbb{Z}_7, +_7)$ is $(\{0\}, +_7)$, the set $\{0\}$ under addition modulo 7. This is a valid subgroup because it satisfies the group axioms. It is closed under the binary operation because $0 +_7 0 = 0$; the operation is associative because $(0 +_7 0) +_7 0 = 0 +_7 (0 +_7 0)$; it contains the identity element 0; and the single element 0 is its own inverse. If you think that $(\{0\}, +_7)$ is a *trivial* subgroup, you will again be pleased to learn that this exactly is what it is sometimes called.

$+_7$	0
0	0

The number of elements in a group $(G, *)$ is called the *order* of the group and is denoted by $|G|$. To me this notation makes sense because $|G|$ is the 'size' of G, just as $|-3|$ is the 'size' of the number -3. Generalizing the above, every cyclic group of prime order has exactly two subgroups, the whole group and the trivial subgroup. Cyclic groups of non-prime order have networks of nested subgroups, the relationships between which can be represented in diagrams like this one for $(\mathbb{Z}_{12}, +_{12})$:

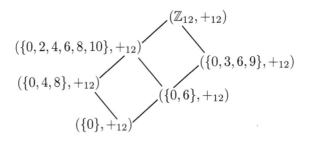

Using generator notation, the labels can be abbreviated.

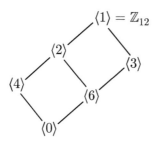

$$\langle 1 \rangle = \mathbb{Z}_{12}$$

What would diagrams look like for $(\mathbb{Z}_7, +_7)$, $(\mathbb{Z}_8, +_8)$ and $(\mathbb{Z}_{60}, +_{60})$?

One thing to remember is that a subgroup must have the same operation as its group. That might sound obvious, but people can get muddled if they do not differentiate groups and sets, a problem compounded by mathematicians omitting the operation when dealing with standard groups: for $(\mathbb{Z}_{12}, +_{12})$, they often write $\mathbb{Z}_{12}$. This abbreviation is perfectly reasonable, and I will use it unless I sense ambiguity. But it can tempt students into saying that $\mathbb{Z}_4$ is a subgroup of $\mathbb{Z}_{12}$. This is not true because although $\{0, 1, 2, 3\} \subseteq \{0, 1, 2, 3, 4, 5, 6, 7, 8, 9, 10, 11\}$, $\mathbb{Z}_4$ has operation $+_4$, not $+_{12}$, and the set $\{0, 1, 2, 3\}$ is not closed under $+_{12}$: for instance, $3 + 2 = 5 \notin \{0, 1, 2, 3\}$.

That said, if you are sensitive to abstract structures then this might give you a 'yeah, but …' feeling. You might want to say that $\mathbb{Z}_4$ *is* a subgroup of $\mathbb{Z}_{12}$ because although the four-element set $\{0, 1, 2, 3\}$ is not closed under $+_{12}$, the four-element set $\{0, 3, 6, 9\}$ is, and $(\{0, 3, 6, 9\}, +12)$ has the same structure as $(\mathbb{Z}_4, +_4)$. If you did think that, good. If not, examine the following tables. These involve different objects but have identical structures. In mathematical terms, their structures are *isomorphic*, so it is accurate to say that $\mathbb{Z}_{12}$ has a subgroup *isomorphic to* $\mathbb{Z}_4$. Isomorphisms are discussed in Chapter 8.

$+_{12}$	0	3	6	9
0	0	3	6	9
3	3	6	9	0
6	6	9	0	3
9	9	0	3	6

$+_4$	0	1	2	3
0	0	1	2	3
1	1	2	3	0
2	2	3	0	1
3	3	0	1	2

6.4 Cyclic subgroups and generators

Did you notice that in the subgroup diagram for $\mathbb{Z}_{12}$, all of the subgroups are cyclic? Each is generated by a single element, including the trivial subgroup $(\{0\}, +_{12})$, which is generated by 0. In general, if a subgroup contains the element g, it must by closure contain $g * g$ and $g * g * g$ and so on. If a subgroup of $\mathbb{Z}_{12}$ contains 3, it must contain $3 +_{12} 3 = 6$ and $3 +_{12} 3 +_{12} 3 = 9$ and $3 +_{12} 3 +_{12} 3 +_{12} 3 = 0$. Similarly, every element of $\mathbb{Z}_{12}$ generates a cyclic subgroup; sometimes this subgroup is the whole of $\mathbb{Z}_{12}$.

Could $\mathbb{Z}_{12}$ also have non-cyclic subgroups? The fact that every element generates a cyclic subgroup does not imply that every subgroup is cyclic—perhaps there also exist subgroups without single generators. We can explore by supposing that a subgroup H of $\mathbb{Z}_{12}$ contains, say, both 3 and 4. It must contain every multiple of 3 and of 4, so $H \supseteq \{3, 6, 9, 0, 4, 8\}$. Is that it? No: H must also contain every combination of these elements. In particular, it must contain $9 +_{12} 4 = 1$. And because it contains the generator 1, it must be the whole group.

Perhaps similar reasoning convinces you that every subgroup of $\mathbb{Z}_{12}$ must be cyclic. But maybe in bigger cyclic groups, weirder things can happen. What do you think? Is that possible? Might 'big' cyclic groups have non-cyclic subgroups? It turns out—as in the theorem quoted in Section 3.5—that the answer is no: every subgroup of every cyclic group is cyclic. To prove this it helps to work with general notation and to consider both generators and *relations*. The group $\mathbb{Z}_{12}$ has a single generator and satisfies the *relation* that adding the generator to itself 12 times gives the identity:

$$\underbrace{1 +_{12} 1 +_{12} 1 +_{12} 1 +_{12} 1 +_{12} 1 +_{12} 1 +_{12} 1 +_{12} 1 +_{12} 1 +_{12} 1 +_{12} 1}_{12} = 0.$$

Specifying generators and relations provides an economical way to represent groups; to standardize across groups, this is usually done in multiplicative notation. The group $\mathbb{Z}_{12}$ can be written $\langle a | a^{12} = e \rangle$, meaning the group generated by a subject to the relation that a^{12} is the identity. In this notation, the table for $\mathbb{Z}_{12}$ appears as below. This strips away links to congruence classes and focuses attention on the abstract structure.

	e	a	a^2	a^3	a^4	a^5	a^6	a^7	a^8	a^9	a^{10}	a^{11}
e	e	a	a^2	a^3	a^4	a^5	a^6	a^7	a^8	a^9	a^{10}	a^{11}
a	a	a^2	a^3	a^4	a^5	a^6	a^7	a^8	a^9	a^{10}	a^{11}	e
a^2	a^2	a^3	a^4	a^5	a^6	a^7	a^8	a^9	a^{10}	a^{11}	e	a
a^3	a^3	a^4	a^5	a^6	a^7	a^8	a^9	a^{10}	a^{11}	e	a	a^2
a^4	a^4	a^5	a^6	a^7	a^8	a^9	a^{10}	a^{11}	e	a	a^2	a^3
a^5	a^5	a^6	a^7	a^8	a^9	a^{10}	a^{11}	e	a	a^2	a^3	a^4
a^6	a^6	a^7	a^8	a^9	a^{10}	a^{11}	e	a	a^2	a^3	a^4	a^5
a^7	a^7	a^8	a^9	a^{10}	a^{11}	e	a	a^2	a^3	a^4	a^5	a^6
a^8	a^8	a^9	a^{10}	a^{11}	e	a	a^2	a^3	a^4	a^5	a^6	a^7
a^9	a^9	a^{10}	a^{11}	e	a	a^2	a^3	a^4	a^5	a^6	a^7	a^8
a^{10}	a^{10}	a^{11}	e	a	a^2	a^3	a^4	a^5	a^6	a^7	a^8	a^9
a^{11}	a^{11}	e	a	a^2	a^3	a^4	a^5	a^6	a^7	a^8	a^9	a^{10}

We can use this notation to prove that every subgroup of a cyclic group is cyclic. If you understand the intuitive arguments, then the proof below might seem unnecessary. But if you like to see how mathematics fits together, you might like it. It relies upon properties of exponents and the *division algorithm* for the integers, which states that if n is an integer and m is a positive integer, then there exist unique integers q and r such that

$$n = mq + r \quad \text{and} \quad 0 \le r < m.$$

For instance, if $n = 27$ and $m = 6$, then there exist unique integers q and r such that

$$27 = 6q + r, \text{ where it happens that } q = 4 \text{ and } r = 3.$$

This familiar knowledge is used in the following theorem and proof. Note that in the proof, the reason for the set-up becomes clear toward the end. So if, at some point, you are not sure what is happening, remember the self-explanation training from Section 3.6 and be ready to read back and forth.

Theorem: Let G be a cyclic group. Then every subgroup of G is cyclic.

Proof: Suppose that G is cyclic and that $G = \langle a \rangle$.

Suppose that H is a subgroup of G.

If $H = \{e\}$ then H is cyclic.

If $H \neq \{e\}$, let m be the smallest positive integer such that $a^m \in H$.

Claim: $H = \langle a^m \rangle$.

Proof of claim:

Let $b \in H \subseteq G$ and suppose that $b = a^n$.

Then $\exists q, r \in \mathbb{Z}$ where $0 \leq r < m$ such that $n = mq + r$.

Thus $b = a^n = a^{mq+r} = (a^m)^q a^r$.

So $a^r = (a^m)^{-q} a^n$.

Now $a^n, a^m \in H$ and H is a subgroup, so $a^r \in H$ by closure.

But $0 \leq r < m$ and m is the smallest positive integer such that $a^m \in H$.

So $r = 0$, meaning that $n = qm$ and $b = a^n = (a^m)^q$.

Thus b is a power of a^m.

So $H = \langle a^m \rangle$.

Hence H is cyclic.

We can also generalize the observation that in $\mathbb{Z}_{12}$, each element generates a cyclic subgroup. This is a big generalization because in every group—not just cyclic groups—each element generates a cyclic subgroup. You might want to check that you believe this for the groups considered so far. What subgroups are generated by each element of D_3, for instance? How would the following theorem and proof capture their properties?

Theorem: Let G be a group and $g \in G$. Then $\langle g \rangle$ is a subgroup of G.

Proof: Let G be a group and let $g \in G$.

    Let $a, b, c \in \langle g \rangle$.

    Then $\exists\, l, m, n \in Z$ such that $a = g^l, b = g^m$ and $c = g^n$. So

- $\langle g \rangle$ is closed under the group operation because
  $ab = g^l g^m = g^{l+m} \in \langle g \rangle$.
- The group operation is associative on $\langle g \rangle$ because
  $(ab)c = (g^l g^m)g^n = g^{l+m}g^n = g^{l+m+n} = g^l g^{m+n} = g^l(g^m g^n) = a(bc)$.
- The group identity element $e \in \langle g \rangle$ because $g^0 = e$.
- Every element in $\langle g \rangle$ has its inverse in $\langle g \rangle$ because $g^{-l} \in \langle g \rangle$ and
  $g^l g^{-l} = g^{-l} g^l = e$.

    Thus $\langle g \rangle$ is a subgroup of G; it is cyclic because it has a single generator.

As usual, it is worth considering choices made in proof presentation. For instance, for the above proof I vacillated on notation. First I denoted the subgroup by $\langle a \rangle$ and elements of $\langle a \rangle$ by b, c, d. Then I decided that I didn't like that a had a different status from b, c and d, so I changed b, c and d to x, y and z. Then I decided that I didn't like having two 'sets' of letters not involving g, so I changed it again. All of these options would be fine—the decision is a matter of communication. Think about whether other changes would clarify the argument for you and, if you fancy a challenge, try writing the above two theorems and proofs in additive notation.

6.5 Theorems about cyclic groups

Most sections of this chapter have started with specific groups and invited you to generalize. But an Abstract Algebra course might present theorems and proofs without much intuitive preamble. You will therefore have to work out how general theorems apply to familiar groups. This can be harder than it sounds, not least because students sometimes try to read a proof before they've fully understood the corresponding theorem. This section will therefore use cyclic groups to provide practice in understanding theorems.

First, something simple. Below you can find a definition (first mentioned in Section 2.5), along with a theorem. Think about what the theorem means. How would you explain it informally? Then think about why it must be true. How can two elements of G be represented for a group of the form $G = \langle a \rangle$? For G to be abelian, what must be true about these elements? Which group properties might be useful for a proof?

Definition: A group $(G, *)$ is **abelian** if and only if $*$ is commutative on G.

Theorem: Suppose that $G = \langle a \rangle$ is a cyclic group. Then G is abelian.

Second, consider the theorem below. This provides good understanding practice because it is a bit of a mouthful but, once you get past that, it is fairly accessible. Note that 'gcd' means 'greatest common divisor', which you might have seen expressed as 'hcf' for 'highest common factor'. With that in mind, try reading the theorem aloud. Then try working out why it is reasonable by applying it to the group $\mathbb{Z}_{12}$. Because the theorem is stated in multiplicative notation, it might be easiest to think of $\mathbb{Z}_{12}$ as $\langle a | a^{12} = e \rangle$, so that each element is of the form a^n.

Theorem: Suppose that $G = \langle a \rangle$ and $|G| = n$. Let $b \in G$ and $b = a^s$. Then b generates a cyclic subgroup of G containing n/d elements, where $d = \gcd\{n, s\}$.

Here is how I would think. For $G = \mathbb{Z}_{12}$, $n = 12$. For $b \in G$ I might pick $b = a^5$, so $s = 5$. Then the theorem says that a^5 generates a cyclic subgroup of G containing n/d elements, where $d = \gcd\{n, s\} = \gcd\{12, 5\} = 1$ because 12 and 5 have no common divisors (factors). That is, 5 generates a subgroup of $\mathbb{Z}_{12}$ containing $12/1 = 12$ elements. That is correct because a^5 generates the whole group. It doesn't best showcase how the theorem works, however, because a^5 generates an improper subgroup. So it is worth trying again, choosing $b = a^3$ or $b = a^4$ or $b = a^8$. Perhaps you can immediately see how that would work. If not, think it through. It would also be worth translating the theorem into the original additive notation for $\mathbb{Z}_{12}$, with elements written as numbers representing congruence classes.

Next, consider the third theorem below.

Theorem: If $G = \langle a \rangle$ and $|G| = n$ then other generators of G are of the form a^r, where r and n are coprime.

Again, try applying this to $\mathbb{Z}_{12}$, checking against your earlier list of generators. Does the theorem reflect what you already know? Then try using it to find generators of $(\mathbb{Z}_{60}, +_{60})$.

This third theorem is a *corollary* of the second, meaning that it follows from it more or less directly. Can you work out why? Try asking what d must be in the second theorem for b to generate the whole group. The third theorem also provides a way to prove the earlier observation that a group $\mathbb{Z}_p$ with p prime has no proper nontrivial subgroups. Can you see why?

Now, the preceding two theorems are about finite cyclic groups. But much reasoning in this chapter applies to every cyclic group. This might seem to make no difference: the cyclic group $\mathbb{Z}_n$ 'cycles around' through n elements. But cyclic groups do not have to be finite. The circular image is not universal, because the definition of cyclic group is not about circles: it is about the group having a single generator. Here it is again, in both multiplicative and additive forms.

Definition: Let G be a group. Then G is cyclic if and only if $\exists g \in G$ such that $G = \langle g \rangle = \{g^n | n \in \mathbb{Z}\}$.

Definition: Let G be an additive group. Then G is cyclic if and only if $\exists g \in G$ such that $G = \langle g \rangle = \{ng | n \in \mathbb{Z}\}$.

These definitions mean that the infinite group $(\mathbb{Z}, +)$ is cyclic: the element 1 is a generator because $\mathbb{Z} = \langle 1 \rangle = \{n1 | n \in \mathbb{Z}\}$. This *feels* different from generators in a finite cyclic group, because starting at the identity and adding the generator 1 doesn't seem to give everything.

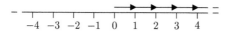

The definition, however, says that an element is a generator if and only if the set containing every integer multiple of that element is the whole group. The 'keep adding the generator' intuition doesn't quite correspond to that because it uses every *positive* integer multiple of the generator, not every integer multiple. In finite cyclic groups this makes little difference because positive multiples alone or positives and negatives together give the whole group.

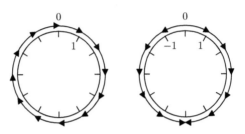

In the infinite cyclic group $(\mathbb{Z}, +)$, it makes a big difference.

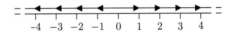

Using circles to think about cyclic groups is therefore intuitively helpful but 'a bit wrong' in that it does not quite work for all cases. Consequently, it would be worth reviewing this chapter to work out which ideas apply to $(\mathbb{Z}, +)$. For instance, does $(\mathbb{Z}, +)$ have alternative generators? Yes: $g = -1$ generates $\mathbb{Z}$, because $\langle -1 \rangle = \{n(-1) | n \in \mathbb{Z}\} = \mathbb{Z}$. But that's it. Every other element generates a proper subgroup of $\mathbb{Z}$. For instance, the element 3 generates the subgroup

$$\langle 3 \rangle = \{n3 | n \in \mathbb{Z}\} = \{\ldots, -6, -3, 0, 3, 6, \ldots\} = 3\mathbb{Z}.$$

Indeed, every subgroup of $(\mathbb{Z}, +)$ takes the form $(n\mathbb{Z}, +)$ because the theorem about subgroups of cyclic groups applies. Which other theorems apply to $(\mathbb{Z}, +)$, and does re-reading with $(\mathbb{Z}, +)$ in mind deepen your understanding?

To conclude, a note on notation: it is important not to mix up $n\mathbb{Z}$ and $\mathbb{Z}_n$. The set $n\mathbb{Z}$ is the infinite set of integer multiples of n, which forms a subgroup of $\mathbb{Z}$ under standard addition. The set $\mathbb{Z}_n$ is the finite set of congruence classes of the integers under addition modulo n. It is a group under $+_n$, but it is *not* a subgroup of $\mathbb{Z}$. It cannot be, because its elements are not integers and $+_n$ is not $+$. If, however, you have noticed that $n\mathbb{Z}$ and $\mathbb{Z}_n$ are closely linked, you will probably enjoy Chapter 7.

6.6 Groups of familiar objects

So far, this chapter has been mostly about cyclic groups, which pop up everywhere—all groups have cyclic subgroups, for instance, because every element of every group generates a cyclic subgroup. But now we will consider other groups, starting with 'ordinary' numbers under multiplication.

Do the integers form a group under multiplication? It looks like they might. The set $\mathbb{Z}$ is closed under multiplication, multiplication is associative because if x, y and z are integers then $(xy)z = x(yz)$, and a multiplicative identity exists—the number 1. But most integers do not have multiplicative inverses: for instance, there is no integer x such that $2x = x2 = 1$. So $(\mathbb{Z}, \times)$ is not a group.

That does, however, suggest an extension. If the integers do not form a group under multiplication, how about the rationals, denoted $\mathbb{Q}$? Rationals are numbers of the form p/q, where $p, q \in \mathbb{Z}$ and $q \neq 0$. The set $\mathbb{Q}$ is closed under multiplication—why? Again multiplication is associative with identity 1. Moreover, p/q has multiplicative inverse q/p, so we also have inverses. Or do we? Not quite, because $0 \in \mathbb{Q}$ ($0 = 0/1$, for instance) but it has no multiplicative inverse. That's a shame. However, the set $\mathbb{Q}\backslash\{0\}$ does form a group under multiplication. Does $\mathbb{Q}$, with 0 included, form a group under addition?

We can ask similar questions about the real numbers, denoted $\mathbb{R}$, and the complex numbers, denoted $\mathbb{C}$. Do these form groups under addition or multiplication, perhaps with 0 excluded? Think about this and you will see that it is not hard to identify groups formed by familiar numbers and operations. However, such groups might not get much airtime in group theory courses because, while they are perfectly good groups, they are not *just* groups. For instance, $\mathbb{Z}$ is not just a group under addition, it is a *ring* under addition and multiplication. This classification more fully captures its structure. Similarly, $\mathbb{R}$ is not just a group under addition, it is a *field* under addition and multiplication. Rings and fields are groups with extra structure—lots of extra structure, in some cases—and will be discussed in Chapter 9.

It is, however, interesting to compare groups of numbers with other groups. For instance, the group $(\mathbb{Z}, +)$ is cyclic. How about $(\mathbb{Q}, +)$ and $(\mathbb{R}, +)$? Are these cyclic? Neither is generated by 1, because integer multiples of 1 give only the integers. But could there be alternative generators? If so, what are they? If not, why not?

We could also think about subgroups. For groups of numbers, this is a big question: $\mathbb{Q}$, $\mathbb{R}$ and $\mathbb{C}$ have more complex structures than $\mathbb{Z}$. Nevertheless, inspired by subgroups of $\mathbb{Z}$, we might ask whether $(\{nx | x \in \mathbb{R}\}, +)$ is always a subgroup of $(\mathbb{R}, +)$. If it is, why? If not, why not? How about $(\{nx + 1 | x \in \mathbb{R}\}, +)$? Does the value of x matter? Do the positive reals form a subgroup of $(\mathbb{R}, +)$ under addition, or perhaps under multiplication? And how about finite subgroups? The group $(\mathbb{Z}, +)$ has no finite subgroups except $(\{0\}, +)$. Does it follow that the groups $(\mathbb{Q}, +)$, $(\mathbb{R}, +)$ and $(\mathbb{C}, +)$ also have no finite subgroups except $(\{0\}, +)$? These groups are 'more infinite' so perhaps any proper nontrivial subgroups must also be 'big'. What do you think?

Either way, multiplicative groups are dramatically different. For instance, $(\{1, -1\}, \times)$ is a (very) finite subgroup of $(\mathbb{Q}\backslash\{0\}, \times)$.

$\times$	1	-1
1	1	-1
-1	-1	1

And this extends elegantly to complex numbers.[2] The table below represents the subgroup $(\{1, i, -1, -i\}, \times)$ of $(\mathbb{C}\backslash\{0\}, \times)$. Do you recognize its structure? What if we rewrite the elements as powers of i?

[2] As noted in Chapter 1, this book includes examples based on complex numbers and matrices. If you are studying in a UK-like system and have not come across these, you can find introductions in A level Further Mathematics textbooks or in reliable online resources.

×	1	i	−1	−i
1	1	i	−1	−i
i	i	−1	−i	1
−1	−1	−i	1	i
−i	−i	1	i	−1

×	1	i	i^2	i^3
1	1	i	i^2	i^3
i	i	i^2	i^3	1
i^2	i^2	i^3	1	i
i^3	i^3	1	i	i^2

If you are familiar with complex numbers, you might notice that $1, i, -1$ and $-i$ are the *fourth roots of unity*: raising each to the power four gives 1. Using an *Argand diagram* of the complex plane, the roots of unity are equally spaced around the *unit circle* (circles again, look). The fifth roots of unity are shown on the right below, where I labelled only some things to avoid clutter. These are perhaps most easily related to the diagram when written in *polar* form $\cos\theta + i\sin\theta$, where θ is the angle anticlockwise from the positive x-axis. In that notation, the fifth roots of unity are

$$1, \quad \cos\tfrac{2\pi}{5} + i\sin\tfrac{2\pi}{5}, \quad \cos\tfrac{4\pi}{5} + i\sin\tfrac{4\pi}{5}, \quad \cos\tfrac{6\pi}{5} + i\sin\tfrac{6\pi}{5},$$
$$\cos\tfrac{8\pi}{5} + i\sin\tfrac{8\pi}{5}.$$

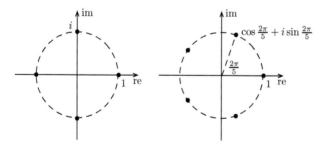

You can check these claims by using trigonometric identities to show, for instance, that $(\cos\tfrac{2\pi}{5} + i\sin\tfrac{2\pi}{5})^5 = 1$, or by recalling that

$r(\cos\theta + i\sin\theta) = re^{i\theta}$ and working with exponents (which show conveniently that the product of two fifth roots of unity is another). Multiplication can be conceptualized as rotating around the unit circle, as illustrated below.

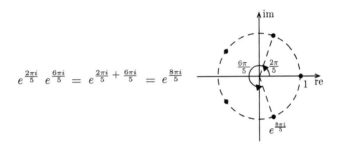

$$e^{\frac{2\pi i}{5}} \ e^{\frac{6\pi i}{5}} \ = \ e^{\frac{2\pi i}{5} + \frac{6\pi i}{5}} \ = \ e^{\frac{8\pi i}{5}}$$

Does this convince you that the fifth roots of unity form a multiplicative group that is structurally identical to $(\mathbb{Z}_5, +_5)$? Generalizing, the set of nth roots of unity is sometimes denoted $U_n = \{z \in \mathbb{C} | z^n = 1\}$, and under multiplication is structurally identical to $(\mathbb{Z}_n, +_n)$; we say that $(U_n, \times)$ is *isomorphic to* $(\mathbb{Z}_n, +_n)$, and write $(U_n, \times) \cong (\mathbb{Z}_n, +_n)$.

Moreover, the entire unit circle forms a group under multiplication. That might sound unnatural because a circle seems like a unified object rather than a set. But a circle is also a set of points. The unit circle is $U = \{z \in \mathbb{C} | |z| = 1\}$; it contains all points of the form $1(\cos\theta + i\sin\theta) = 1e^{i\theta}$. And U is closed under multiplication, has associative multiplication inherited from $\mathbb{C}$, contains the multiplicative identity 1, and contains multiplicative inverses: $z = 1(\cos\theta + i\sin\theta) = 1e^{i\theta}$ has inverse $z^{-1} = 1(\cos(-\theta) + i\sin(-\theta)) = 1e^{i(-\theta)}$. An interesting isomorphism involving the group $(U, \times)$ is discussed in Chapter 8.

And the links do not stop there. Rotations of the whole plane about $(0,0)$ form a group under composition: composing two rotations gives another; composition is associative—think literally about rotations or see Section 5.4 on function composition; rotation through 0 is the identity; and the inverse of rotation through θ is rotation through $-\theta$. Moreover, you might know from Linear Algebra or recall from

Section 5.6 that a rotation through θ about $(0,0)$ can be represented using the matrix

$$\begin{pmatrix} \cos\theta & -\sin\theta \\ \sin\theta & \cos\theta \end{pmatrix}.$$

Composing rotations corresponds to multiplying matrices, so matrices of this form constitute a group under multiplication. And rotation composition is commutative, so multiplication of rotation matrices is commutative, even though matrix multiplication in general is not. You could use matrix calculations to establish this directly.

Of course, rotations correspond to only a restricted class of 2×2 matrices. Do they form a subgroup of $M_{2\times 2}(\mathbb{R})$? In fact they cannot, because $M_{2\times 2}(\mathbb{R})$ is not a group so cannot have subgroups. But the *invertible* or *nonsingular* $n \times n$ matrices over $\mathbb{R}$ do form a group under matrix multiplication. This group is called $GL(n, \mathbb{R})$, the general linear group of degree n over $\mathbb{R}$. The rotation matrices do form a subgroup of $GL(2, \mathbb{R})$. And rotations considered differently form subgroups of dihedral groups.

6.7 The dihedral group D_3

The *dihedral group* D_3 is the group of symmetries of an equilateral triangle, mentioned in Section 1.3 and discussed in Section 5.7. The group operation is composition—symmetries are combined by performing one then another.

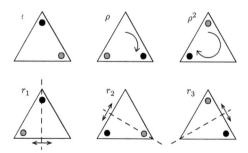

∘	e	ρ	ρ^2	r_1	r_2	r_3
e	e	ρ	ρ^2	r_1	r_2	r_3
ρ	ρ	ρ^2	e	r_2	r_3	r_1
ρ^2	ρ^2	e	ρ	r_3	r_1	r_2
r_1	r_1	r_3	r_2	e	ρ^2	ρ
r_2	r_2	r_1	r_3	ρ	e	ρ^2
r_3	r_3	r_2	r_1	ρ^2	ρ	e

Now, D_3 has six elements. Is it isomorphic—structurally identical—to $\mathbb{Z}_6$? If you trust intuition based on group tables, you might feel that it is not, because the structures are different: the table for $\mathbb{Z}_6$ has diagonal 'stripes', whereas this table has 'blocks'. But could reordering the elements reveal hidden stripes? One way to answer is to consider group properties. Every cyclic group is commutative. And D_3 is not: for instance, $\rho r_1 \neq r_1 \rho$. Reordering cannot make non-commutativity 'go away' because it does not change the result of any composition. So D_3 is not cyclic.

Another approach is to consider the definition, which says that a cyclic group is generated by a single element. In D_3, the identity e generates the single-element set $\{e\}$. The rotations ρ and ρ^2 each generate the subgroup $\{e, \rho, \rho^2\}$. What does a reflection generate? Repeatedly performing any reflection gives the reflection then the identity then the reflection then the identity, and so on. So each reflection generates a two-element subgroup. Thus D_3 is not generated by a single element, so is not cyclic.

That raises a question, however. If a single element is not enough to generate D_3, how many are needed? What do you think? Can two elements be combined in different ways to give every group element? Does it matter which two? I recommend that before reading on, you make a triangle and experiment.

The two rotations ρ and ρ^2 do not generate D_3—each is in the proper subgroup generated by the other. How about two reflections? Composing any two reflections flips the triangle over then back again, so the result must be a rotation. For instance, composing r_1 and r_2 gives $r_1 r_2 = \rho^2$ and $r_2 r_1 = \rho$, so $\langle r_1, r_2 \rangle$ contains at least the four elements r_1, r_2, ρ^2, and ρ. Does it also contain others? You could check directly, or we could reason that $\langle r_1, r_2 \rangle$ must be a subgroup of D_3 (you might like to think about why). If you thought about possible subgroup orders in Sections 3.5 or 6.2, you might know or intuit that in a finite group, the order of a subgroup must divide the order of the group (more on that in Section 7.7). So four elements is too many for a proper subgroup of a six-element group, and it must be that $\langle r_1, r_2 \rangle = D_3$. Do two distinct reflections always generate D_3, or is $\langle r_1, r_2 \rangle$ special? For me it seems clear that all reflections are 'equivalent' in how they interact, so that any two will do. If that does not seem clear to you, explore systematically.

Now, although D_3 is generated by two distinct reflections, mathematicians more usually conceive of it as generated by a reflection and a rotation. Consider the rotation ρ and reflection r (renamed as we only need one) below. These generate D_3 because combining the reflection with powers of the rotation gives every possible symmetry.

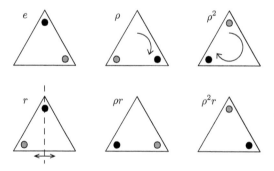

This means that the table can be rewritten using generator-based names. Nothing changes in the symmetry combinations, so the results correspond directly to the earlier notation. Inspect the table below carefully. Which results are 'obvious' and which are not?

∘	e	ρ	ρ^2	r	ρr	$\rho^2 r$
e	e	ρ	ρ^2	r	ρr	$\rho^2 r$
ρ	ρ	ρ^2	e	ρr	$\rho^2 r$	r
ρ^2	ρ^2	e	ρ	$\rho^2 r$	r	ρr
r	r	$\rho^2 r$	ρr	e	ρ^2	ρ
ρr	ρr	r	$\rho^2 r$	ρ	e	ρ^2
$\rho^2 r$	$\rho^2 r$	ρr	r	ρ^2	ρ	e

Many results are obvious: it is not surprising that $(\rho^2)(r) = \rho^2 r$. Others follow directly from $\rho^3 = e$ or $r^2 = e$: using associativity, $(\rho^2)(\rho r) = (\rho^3)r = er = r$. Check and you will see that this accounts for every outcome in the top half of the table. But in the bottom half, some unaccounted-for relationships appear:

$$r\rho = \rho^2 r \quad r\rho^2 = \rho r \quad r\rho r = \rho^2 \quad r\rho^2 r = \rho \quad \rho r\rho = r \quad \rho r\rho^2 = \rho^2 r$$
$$\rho r\rho r = e \quad \rho r\rho^2 r = \rho^2 \quad \rho^2 r\rho = \rho r \quad \rho^2 r\rho^2 = r \quad \rho^2 r\rho r = \rho \quad \rho^2 r\rho^2 r = e.$$

These equations must be valid because all I did was rewrite the table using new labels. Nevertheless, I feel better when I have solidified my intuition by performing some checks. For instance, for the first equation, triangle manipulations look like this.

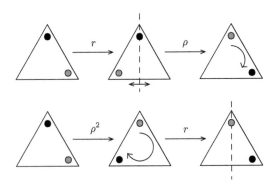

The outcomes match, which is good. But the number of equations is worrying. Does representing D_3 using generators and relations require them all? Fortunately not: it requires only one. Mathematicians often use $r\rho r = \rho^2$, commonly rewriting this as $r\rho r = \rho^{-1}$ because $\rho^2 = \rho^{-1}$ and this formulation generalizes better to other dihedral groups (see Section 6.8). Can you see why only one equation is needed? All are equivalent due to algebraic manipulation and the more obvious relations $\rho^3 = e$ and $r^2 = e$. For instance, remembering that left and right composition must be distinguished because D_3 is not commutative,

$$r\rho = \rho^2 r \;\Leftrightarrow\; r\rho r = \rho^2 r^2 \;\Leftrightarrow\; r\rho r = \rho^2 \;\Leftrightarrow\; r\rho r = \rho^{-1};$$
$$\rho r \rho r = e \;\Leftrightarrow\; \rho^{-1}\rho r \rho r = \rho^{-1} \;\Leftrightarrow\; r\rho r = \rho^{-1}.$$

Try some more and you will see that they also yield nothing new. The relation linking the two generators is sometimes referred to as the *dihedral relation*, so that D_3 can be expressed in either of the forms below.

$$D_3 = \langle \rho, r \,|\, \rho^3 = r^2 = e,\; r\rho r = \rho^{-1} \rangle;$$
$$D_3 = \langle \rho, r \,|\, \rho^3 = r^2 = (\rho r)^2 = e \rangle.$$

Now, what do you think of this notation compared with the original $e, \rho, \rho^2, r_1, r_2, r_3$? I think that there is no obvious best choice. Generator-based notation is 'clean' in that it uses only two symbols and highlights underlying relationships. But the original notation highlights patterns in the table and types of symmetry (ρr involves one of each symbol but is a reflection, for instance). I will therefore stick with the original for later chapters. But Abstract Algebra involves recasting familiar knowledge in new ways, and your lecturer might have different priorities.

To conclude discussion of D_3, note that although D_3 is not cyclic, all of its proper subgroups are: each is trivial or isomorphic to either $\mathbb{Z}_2$ or $\mathbb{Z}_3$. This shows that the converse of an earlier theorem is not true:

$$G \text{ cyclic} \;\Rightarrow\; \text{every proper subgroup of } G \text{ cyclic} \quad \text{TRUE}$$
$$G \text{ cyclic} \;\Leftarrow\; \text{every proper subgroup of } G \text{ cyclic} \quad \text{FALSE}$$

In fact, D_3 is the smallest non-abelian group. Can you work out why smaller groups must be abelian? I will ask again at the end of the chapter.

6.8 More symmetry groups

The obvious group to investigate next is D_4, the group of symmetries of a square.

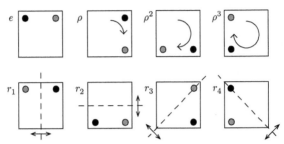

Does the information about D_3 generalize to D_4? Maybe, because D_3 and D_4 are similarly constructed. But D_4 is bigger, so there is 'room' for more internal structure. What do you think? What are the subgroups of D_4, and are they all cyclic? Can two elements generate D_4?

Like D_3, the group D_4 has a two-element subgroup generated by each reflection and a subgroup $\{e, \rho, \rho^2, \rho^3\}$ comprising the rotations and the identity. This time, the rotation subgroup itself has a nontrivial proper subgroup $\{e, \rho^2\}$. Can you see this in the table and understand it by imagining transformations?

∘	e	ρ	ρ^2	ρ^3	r_1	r_2	r_3	r_4
e	e	ρ	ρ^2	ρ^3	r_1	r_2	r_3	r_4
ρ	ρ	ρ^2	ρ^3	e	r_3	r_4	r_2	r_1
ρ^2	ρ^2	ρ^3	e	ρ	r_2	r_1	r_4	r_3
ρ^3	ρ^3	e	ρ	ρ^2	r_4	r_3	r_1	r_2
r_1	r_1	r_4	r_2	r_3	e	ρ^2	ρ^3	ρ
r_2	r_2	r_3	r_1	r_4	ρ^2	e	ρ	ρ^3
r_3	r_3	r_1	r_4	r_2	ρ	ρ^3	e	ρ^2
r_4	r_4	r_2	r_3	r_1	ρ^3	ρ	ρ^2	e

The group D_4 also has a four-element subgroup $\{e, \rho^2, r_3, r_4\}$, highlighted below in the table and a mini-table. Is this subgroup cyclic? No: each non-identity element is self-inverse so it cannot have a single generator.

$\circ$	e	ρ	ρ^2	ρ^3	r_1	r_2	r_3	r_4
e	e	ρ	ρ^2	ρ^3	r_1	r_2	r_3	r_4
ρ	ρ	ρ^2	ρ^3	e	r_3	r_4	r_2	r_1
ρ^2	ρ^2	ρ^3	e	ρ	r_2	r_1	r_4	r_3
ρ^3	ρ^3	e	ρ	ρ^2	r_4	r_3	r_1	r_2
r_1	r_1	r_4	r_2	r_3	e	ρ^2	ρ^3	ρ
r_2	r_2	r_3	r_1	r_4	ρ^2	e	ρ	ρ^3
r_3	r_3	r_1	r_4	r_2	ρ	ρ^3	e	ρ^2
r_4	r_4	r_2	r_3	r_1	ρ^3	ρ	ρ^2	e

$\circ$	e	ρ^2	r_3	r_4
e	e	ρ^2	r_3	r_4
ρ^2	ρ^2	e	r_4	r_3
r_3	r_3	r_4	e	ρ^2
r_4	r_4	r_3	ρ^2	e

Thus D_4 is unlike D_3 in having proper subgroups that are not cyclic. The full set of subgroups of D_4 is represented below.

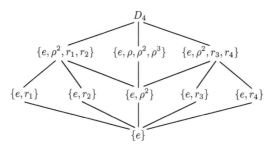

Now, r_3 and r_4 'go together' in that they are diagonal reflections that appear in $\{e, \rho^2, r_3, r_4\}$. But their inelegant positioning in the main table suggests that the elements of D_4 could be labelled in a more structurally natural way. As for D_3, a single rotation and reflection generate D_4.

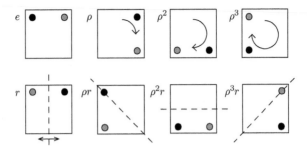

Writing the table using the original notation but in the order equivalent to $r, \rho r, \rho^2 r, \rho^3 r$ reveals diagonal 'stripes' of elements in all four table 'blocks'.

∘	e	ρ	ρ^2	ρ^3	r_1	r_4	r_2	r_3
e	e	ρ	ρ^2	ρ^3	r_1	r_4	r_2	r_3
ρ	ρ	ρ^2	ρ^3	e	r_3	r_1	r_4	r_2
ρ^2	ρ^2	ρ^3	e	ρ	r_2	r_3	r_1	r_4
ρ^3	ρ^3	e	ρ	ρ^2	r_4	r_2	r_3	r_1
r_1	r_1	r_4	r_2	r_3	e	ρ	ρ^2	ρ^3
r_4	r_4	r_2	r_3	r_1	ρ^3	e	ρ	ρ^2
r_2	r_2	r_3	r_1	r_4	ρ^2	ρ^3	e	ρ
r_3	r_3	r_1	r_4	r_2	ρ	ρ^2	ρ^3	e

Also, D_4 can be represented using generators and relations with only the obvious difference from D_3 (what is that difference?):

$$D_4 = \langle \rho, r \mid \rho^4 = r^2 = e,\ r\rho r = \rho^{-1} \rangle;$$
$$D_4 = \langle \rho, r \mid \rho^4 = r^2 = (\rho r)^2 = e \rangle.$$

Do you see why the relation $r\rho r = \rho^{-1}$ applies in all dihedral groups? Why would performing the generating reflection then the rotation then the reflection always gives the inverse of the rotation? I think this is easier to visualize in polygons with more edges—see the regular hexagons below. How does it relate to the fact that the stripes in the table blocks do not all 'go the same way'? And, for the square, hexagon and other polygons, which rotations can and cannot be generators?

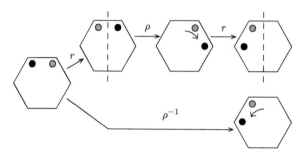

Next, what is D_5 like? How could we denote its elements, and can you write its table in multiple ways? What are its subgroups? Think carefully here. The group D_5 is bigger than D_4, so it has room for more internal structure. But a square has two different 'types' of reflection: those with vertex-to-vertex axes and those with mid-edge-to-mid-edge axes. A regular pentagon does not: like a triangle, its reflections all have vertex-to-mid-edge axes. Moreover, its subgroups must each have order $1, 2, 5$ or 10. Can you list them all?

You can carry on, of course, thinking about D_6, D_7, D_8 and so on. You might also think about symmetry groups for shapes other than regular polygons. For instance, consider a non-square rectangle.

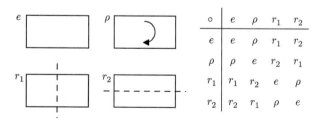

∘	e	ρ	r_1	r_2
e	e	ρ	r_1	r_2
ρ	ρ	e	r_2	r_1
r_1	r_1	r_2	e	ρ
r_2	r_2	r_1	ρ	e

This group is known as the Klein four-group and is often denoted V (the German for 'four' is 'vier'). A group isomorphic to V appears as a subgroup of D_4—which is it? And what other shapes share this symmetry group? A rhombus? A parallelogram that is not a rhombus? A trapezium? If their symmetry groups are not V, what are they?

Finally, many other figures have symmetries. As a child, you might have been asked about symmetry of upper-case letters. What are the symmetry groups of A, F and N? Do any upper-case letters have other symmetry groups? Is there a letter with symmetry group isomorphic to $\mathbb{Z}_4$? If not, can you construct a shape with symmetry group $\mathbb{Z}_4$? How about with symmetry group $\mathbb{Z}_3$ or $\mathbb{Z}_5$? And why stop at two dimensions? What progress can you make in investigating the symmetry group of a cube?

6.9 Permutation groups

The final groups we will explore are groups of *permutations*. As in Section 5.6, a permutation of a set S is a rearrangement of its elements or, more technically, a bijection from S to itself. For $S = \{1, 2, 3, 4\}$, three permutations in 'double-decker' notation appear below. The first sends 1 to 2, 2 to 3, 3 to 4 and 4 to 1; the second swaps 1 and 2 and swaps 3 and 4; the third swaps 2 and 4 but does not move 1 or 3.

$$\begin{pmatrix} 1 & 2 & 3 & 4 \\ 2 & 3 & 4 & 1 \end{pmatrix} \quad \begin{pmatrix} 1 & 2 & 3 & 4 \\ 2 & 1 & 4 & 3 \end{pmatrix} \quad \begin{pmatrix} 1 & 2 & 3 & 4 \\ 1 & 4 & 3 & 2 \end{pmatrix}$$

In the alternative *tuple* notation, the first two of these would be written (1234) and (12)(34). Tuple notation is thus more economical, but it raises representational subtleties: the third permutation could be written (1)(24)(3), but mathematicians write simply (24), assuming that elements not mentioned stay where they are. They could also write (42), because this has the same effect. By convention, though, the lowest number in each bracket is listed first.

Because permutations of a set are functions on that set, any two can be combined via composition, and it is common to write the operation performed first on the right. Below are various representations of permutation composition. When reading the tuple notation, my thinking about the left-hand side of the equality goes like this. I start with (1234), the rightmost permutation. I read 1 goes to 2, then nothing happens to it in the middle bracket, then it goes back to 1 again in the leftmost bracket. So overall it stays where it is and does not appear on the the right-hand side of the equality. Analogously, 2 goes to 3, then 3 goes to 4, then stays where it is. So overall 2 goes to 4. And so on.

$$\begin{pmatrix} 1 & 2 & 3 & 4 \\ 2 & 1 & 4 & 3 \end{pmatrix} \begin{pmatrix} 1 & 2 & 3 & 4 \\ 2 & 3 & 4 & 1 \end{pmatrix} = \begin{pmatrix} 1 & 2 & 3 & 4 \\ 1 & 4 & 3 & 2 \end{pmatrix}$$

$$(12)(34)(1234) = (24)$$

Now, as a beginning undergraduate I spent quite some time on permutation calculations. I felt like I was learning something, and indeed I was learning about a new operation on a new kind of object. But my attention was caught up in the detail—I did not manage the compression necessary to think about permutations as elements of groups, or about how groups of permutations compare with other groups. So here I will assume that your course will provide calculation practice and focus instead on group structure.

First, the full set of permutations of any set S forms a group under composition. Closure holds because composing two permutations gives another. Composition is associative because permutations are functions. The identity permutation leaves every element where it is (we might just write 'e' in tuple notation). And every permutation has an inverse sending each element back to where it came from. The group of all permutations of a set with n elements is denoted by S_n. Its elements are usually written with numbers as above, assuming that $S = \{1, 2, 3, \ldots, n\}$. The group S_n is called the *symmetric group of degree n*. Note: *symmetric* group, not *symmetry* group. That name is different enough to be annoying if symmetric groups and symmetry groups are the same, and similar enough to be annoying if they are different. To work out which it is, we will look at specific symmetric groups, starting small.

The symmetric group S_1 contains all permutations of the one-element set {1}. So it too has one element: the identity permutation, which leaves the 1 where it is. Note that experience with degenerate cases does not necessarily make them seem less weird, and that the set {1} is a set containing one element, whereas the set S_1 is a set containing one function on that element: the objects are different. The symmetric group S_2 of permutations of {1, 2} has two elements: the identity and the permutation that swaps 1 and 2. A table shows that S_2 is isomorphic to $\mathbb{Z}_2$.

$\circ$	e	(12)
e	e	(12)
(12)	(12)	e

$+_2$	0	1
0	0	1
1	1	0

How many elements does S_3 have? It must be 6 because there are $3! = 6$ bijections from a three-element set to itself. Here is a list.

$$e \quad (12) \quad (13) \quad (23) \quad (123) \quad (132)$$

These permutations form a familiar structure: S_3 is isomorphic to D_3, where labelling an equilateral triangle's vertices with numbers provides permutation names for D_3's elements. We know about D_3, so I will not add the table, but you might want to write one.

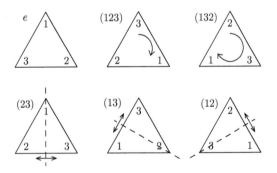

Does it follow that symmetric groups and symmetry groups are the same? Is S_4 isomorphic to D_4? No: it cannot be, because S_4 has $4! = 24$ elements and D_4 has only eight. The elements of S_4 are listed below. Check that the list contains no duplicates, then imagine or make a square with numbered vertices—which permutations correspond to elements of D_4?

e
(12)   (13)   (14)   (23)   (24)   (34)
(123)   (132)   (124)   (142)   (134)   (143)   (234)   (243)
(1234)   (1243)   (1324)   (1342)   (1423)   (1432)
(12)(34)   (13)(24)   (14)(23)

If you try that seriously, you will understand why S_4 has a subgroup isomorphic to D_4. For instance, the permutation (12)(34) corresponds to the reflection shown below, and thus to an element of D_4. But the permutation (12) requires 'twisting' the square, which does not correspond to an element of D_4. With that in mind, try again—which elements of S_4 correspond to elements of D_4? What twists or other disallowed moves would be required for those that do not? Are some of these twists the same twist in disguise?

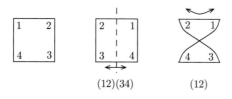

(12)(34)   (12)

We will return to this in Section 7.6. In the meantime, these observations link to the name *symmetric group*. In a square, not all vertices relate to one another in the same way. Under any symmetry of the square, vertex 2 remains adjacent to vertices 1 and 3; no symmetry makes it adjacent to vertex 4. This means that in D_4, there is asymmetry in the way that the numbered vertices can be permuted. In the symmetric group, this asymmetry disappears: S_4 is maximally symmetric in that it treats each element of $S = \{1, 2, 3, 4\}$ in exactly the same way in relation to all the others.

The group S_4 has a subgroup isomorphic to D_4 because the elements of D_4 are all *isometries*—distance-preserving transformations (see Section 5.5)—and the property 'being an isometry' interacts favourably with the group axioms: composing two isometries gives another, associativity is inherited, the identity is an isometry, and every isometry has an inverse isometry. This way of identifying subgroups generalizes to any property that interacts similarly favourably with the group axioms. One such property is fixing elements or sets of elements. For instance, consider the subset of S_4 containing all permutations that fix the number 1. Composing two such permutations gives another, associativity is inherited, the identity fixes 1, and every permutation that fixes 1 has an inverse that fixes 1. This subset is therefore a subgroup; it comprises all permutations that permute the three numbers 2, 3 and 4 so it is isomorphic to D_3.

Another way to identify subgroups is via generators. As in earlier sections, every element or set of elements generates a subgroup (note that we have not proved this—how would you go about it?). What subgroups are generated by individual elements of S_4? How about by pairs or triples of elements? Which combinations generate the whole of S_4? Investigate and you will find that for a group of this size, this is a complex question. It is therefore useful to have systematic ways to think about relationships between permutations, and one way to do this is to consider *transpositions*. A transposition is a two-element permutation like (12), and transpositions combine in useful ways to generate other permutations. For instance, remembering that the permutation on the right acts first,

$$(13)(12) = (123) \text{ and } (14)(13)(12) = (1234).$$

Can you generalize from this to convince yourself that every permutation can be written by composing transpositions? Can some transpositions be made by combining other transpositions? What minimal set of transpositions will generate S_4? Your course might consider that by defining an *even* permutation as the product of an even number of transpositions and an *odd* permutation as the product of an odd number of transpositions. Of course, that makes sense only if all permutations are even or odd but not both. Is that the case? Why? Do the even permutations form a subgroup of S_4, like the even numbers form a subgroup of $(\mathbb{Z}, +)$? And how does all of this generalize to bigger symmetric groups like S_5 or S_6? Chapter 7 will pick up some of these points.

To conclude this section, a point to remember and an observation. The point to remember is that groups of permutations are *not* groups of numbers. They might be represented using numbers, but the elements of permutation groups are *permutations* of these numbers—they are *functions*. Abstraction can help here, I think. For permutations of a set S, the order of S matters—there are more permutations of a set with four elements than there are of a set with three. But the actual elements do not matter: permutations of $\{\alpha, \beta, \gamma, \delta\}$ exactly match those of $\{1, 2, 3, 4\}$. The numbers notation is standard, but bear this in mind.

The observation is that every finite group is isomorphic to a group of permutations. This is because in any possible group table, each row (and column) contains each group element exactly once, so each row (and column) is a permutation of the group elements. Thus every finite group based on a set G is isomorphic to a subgroup of S_G because it is a group formed by a subset of all possible permutations. That observation is very abstract, so it is worth some thought.

6.10 Identifying and defining subgroups

This chapter has considered various specific groups and subgroups. But how can we identify subgroups in general? In small groups, this might be fairly easy. In larger groups, it is more challenging. But we can make progress by extending ideas raised so far. For instance, some properties interact favourably with the group axioms. Because associativity is always inherited from a main group, a property determines a subgroup if it delimits a subset that is closed under the group operation and that includes the identity and all of its elements' inverses. This is a very general idea.

- The subset of S_4 comprising all permutations that fix the number 1 is closed under composition and contains the identity and all of its elements' inverses.
- The subset of S_4 comprising all even permutations is closed under composition and contains the identity and all of its elements' inverses.
- The subset of $(\mathbb{Z}, +)$ comprising all multiples of 3 is closed under composition and contains the identity and all of its elements' inverses.
- The subset of D_4 generated by the element ρ is closed under composition and contains the identity and all of its elements' inverses.

- The subset of $(\mathbb{C}\backslash\{0\}, \times)$ generated by $e^{\frac{2\pi i}{5}}$ is closed under composition and contains the identity and all of its elements' inverses.

Other properties behave similarly. For instance, the *centre* of a group always forms a subgroup, where the centre is defined as below.

Definition: The **centre** of a group G is the set $\{x \in G | \forall g \in G, xg = gx\}$.

In words, the centre of G is the set of all elements that commute with everything in G. In abelian groups, the centre is uninteresting because every element commutes with everything in G so the centre is the whole group. In non-abelian groups it can be interesting, because overall non-commutativity can accommodate specific elements that do commute with everything. For instance, the centre of D_4 contains the identity e, because $eg = ge$ for every $g \in D_4$. Does it contain other elements too? Check in the table below and you will find that the centre of D_4 is $\{e, \rho^2\}$. Can you find elements in the centre of D_3? How about in the centre of $GL(2, \mathbb{R})$?

$\circ$	e	ρ	ρ^2	ρ^3	r_1	r_4	r_2	r_3
e	e	ρ	ρ^2	ρ^3	r_1	r_4	r_2	r_3
ρ	ρ	ρ^2	ρ^3	e	r_3	r_1	r_4	r_2
ρ^2	ρ^2	ρ^3	e	ρ	r_2	r_3	r_1	r_4
ρ^3	ρ^3	e	ρ	ρ^2	r_4	r_2	r_3	r_1
r_1	r_1	r_4	r_2	r_3	e	ρ	ρ^2	ρ^3
r_4	r_4	r_2	r_3	r_1	ρ^3	e	ρ	ρ^2
r_2	r_2	r_3	r_1	r_4	ρ^2	ρ^3	e	ρ
r_3	r_3	r_1	r_4	r_2	ρ	ρ^2	ρ^3	e

Now, I claimed that the centre is always a subgroup. Why is that? Associativity is inherited, and the identity is always in the centre because it always commutes with everything. So the interesting axioms involve closure and inverses. Checking closure requires checking that for every x and y in the centre of G, xy is in the centre of G. That means checking that for all $g \in G$, $(xy)g = g(xy)$. Why must that be true? We can establish it using the premise that x and y are in the centre.

$$
\begin{aligned}
(xy)g &= x(yg) \quad \text{by associativity} \\
&= x(gy) \quad \text{because } yg = gy \\
&= (xg)y \quad \text{by associativity} \\
&= (gx)y \quad \text{because } xg = gx \\
&= g(xy) \quad \text{by associativity.}
\end{aligned}
$$

How about inverses? To confirm that the centre contains all of its elements' inverses, we need to check that for every x in the centre of G, x^{-1} is in the centre of G. That requires checking that for all x in the centre of G and for all $g \in G$, $x^{-1}g = gx^{-1}$. As usual, it helps to start with what we know. If x is in the centre of G, we know that $xg = gx$. Strategic multiplying then gives the required result.

$$
\begin{aligned}
& xg = gx \\
\Rightarrow \ & x^{-1}xg = x^{-1}gx \quad \text{multiplying on the left by } x^{-1} \\
\Rightarrow \ & g = x^{-1}gx \quad \text{because } x^{-1}x = e \\
\Rightarrow \ & gx^{-1} = x^{-1}gxx^{-1} \quad \text{multiplying on the right by } x^{-1} \\
\Rightarrow \ & gx^{-1} = x^{-1}g \quad \text{because } xx^{-1} = e.
\end{aligned}
$$

How would you write a proof that the centre of a group is always a subgroup? Which parts of the above explanation would you keep, and which would you abbreviate or cut?

Now, the information in this section makes it possible to identify some subgroups in various groups. But theory building is also important, so it is useful to examine general criteria guaranteeing that a subset is a subgroup. *Subgroup* can be defined as below (using an explicit operation).

Definition: Let $(G, *)$ be a group and $H \subseteq G$. Then $(H, *)$ is a subgroup of $(G, *)$ if and only if $(H, *)$ is itself a group.

Recalling that associativity is always inherited, a subgroup must therefore satisfy the remaining three group axioms.

Definition: Let $(G, *)$ be a group and $H \subseteq G$. Then $(H, *)$ is a subgroup of $(G, *)$ if and only if:

Closure $\forall a, b \in H, a * b \in H$;

Identity the identity $e \in H$;

Inverses $\forall a \in H, a^{-1} \in H$.

This definition is perfectly adequate for proving that a subset is a subgroup. But mathematicians like their criteria minimal, and the following theorem provides a way to check that a subset is a subgroup using just one criterion.

Theorem: Suppose that $(G, *)$ is a group and $\emptyset \neq H \subseteq G$.

Then $(H, *)$ is a subgroup of $(G, *)$ if and only if $\forall a, b \in H$, $a * b^{-1} \in H$.

This, in my view, is less intuitive because it does not relate directly to ideas like closure, identities and inverses. But it is economical, for instance permitting the following proof that $(3\mathbb{Z}, +)$ is a subgroup of $(\mathbb{Z}, +)$.

Claim: $(3\mathbb{Z}, +)$ is a subgroup of $(\mathbb{Z}, +)$.

Proof: Let $a, b \in 3\mathbb{Z}$, so $\exists n_1, n_2 \in \mathbb{Z}$ such that $a = 3n_1$ and $b = 3n_2$.

Then $b^{-1} = -3n_2$.

So $a + b^{-1} = 3n_1 + (-3n_2) = 3(n_1 + (-n_2)) \in 3\mathbb{Z}$.

So $(3\mathbb{Z}, +)$ is a subgroup of $(\mathbb{Z}, +)$.

The general theorem is proved below, with the proof split into two parts because this is an *if and only if* theorem. One direction, labelled '$\Rightarrow$:', assumes that H is a subgroup and proves that the criterion is satisfied. The other, labelled '$\Leftarrow$:', assumes the criterion and proves that H is a subgroup. Which is harder and which is easier, and why? As you read the proof, remember the self-explanation training from Section 3.6.

Theorem: Suppose that $(G, *)$ is a group and $\emptyset \neq H \subseteq G$.

Then $(H, *)$ is a subgroup of $(G, *)$ if and only if $\forall a, b \in H$, $a * b^{-1} \in H$.

Proof: $\Rightarrow$: Suppose that H is a subgroup and that $a, b \in H$.

Then $b^{-1} \in H$ by the inverse criterion in the subgroup definition.

So $a * b^{-1} \in H$ because H is closed under $*$.

$\Leftarrow$: Assume that $\forall a, b \in H$, $a * b^{-1} \in H$.

In particular, $\forall a \in H$, $a * a^{-1} \in H$.

But $a * a^{-1} = e$ so H contains the identity.

Because $e \in H$, it follows that $\forall b \in H$, $e * b^{-1} = b^{-1} \in H$.

Thus H contains all of its elements' inverses.

Finally, suppose that $a, b \in H$.

Then $b^{-1} \in H$ because H contains all of its elements' inverses.

So, by the condition, $a * (b^{-1})^{-1} = a * b \in H$.

Thus H is closed under the group operation.

So H is a subgroup of G.

To conclude this section, you might like to note that for any group G, the intersection of any two subgroups is a subgroup under the group operation. Why is that? How does it play out for some of the groups discussed in this chapter? Can you prove it using the subgroup definition or the theorem above?

6.11 Small groups

This final section considers groups with small numbers of elements. These are often studied in depth, for at least two reasons. First, they can be understood as building blocks for larger, more complex groups. Second, they allow us to understand how the axioms restrict group structures.

We will start with groups of order four. The cyclic group of order four can be understood as $(\mathbb{Z}_4, +_4)$ or as $U_4 = (\{1, i, -1, -i\}, \times)$. These have different elements and operations but are *isomorphic*: they have identical structures.

$+_4$	0	1	2	3
0	0	1	2	3
1	1	2	3	0
2	2	3	0	1
3	3	0	1	2

$\times$	1	i	-1	$-i$
1	1	i	-1	$-i$
i	i	-1	$-i$	1
-1	-1	$-i$	1	i
$-i$	$-i$	1	i	-1

Do there exist groups of order four that are not cyclic? We have seen one—where? We can explore the general question by attempting to construct group tables using four elements. Because we do not know what structures are possible, I will denote the operation by $*$, the identity by e, and the remaining three elements by a, b and c. The identity property then forces the first row and column.

$*$	e	a	b	c
e	e	a	b	c
a	a			
b	b			
c	c			

What shall we fill in for $a * a$? Not a because the inverse property forces each element to appear once in each row and column (see Section 5.3). But we could use the 'next' element, b, as on the left below. What then goes in the remaining second-row cells? The inverse property means it must be c and e, which must also appear once in each column. Thus the only option is that shown on the right. Completing the table via similar reasoning gives, again, the cyclic group of order 4.

*	e	a	b	c
e	e	a	b	c
a	a	b		
b	b			
c	c			

*	e	a	b	c
e	e	a	b	c
a	a	b	c	e
b	b			
c	c			

However, alternative choices are possible: we could try a different element in the $a * a$ cell. Using c and reasoning about rows and columns leads to the table on the right below. Do check—doing so is quick so this is not the time to just believe me.

*	e	a	b	c
e	e	a	b	c
a	a	c		
b	b			
c	c			

*	e	a	b	c
e	e	a	b	c
a	a	c	e	b
b	b	e	c	a
c	c	b	a	e

Now come the subtleties. This looks different from the cyclic group table—it doesn't have diagonal 'stripes'. But could it be the cyclic group in disguise? It turns out that it could. Switching the order of b and c while respecting the outcomes of the operation gives the rewritten table below.

*	e	a	c	b
e	e	a	c	b
a	a	c	b	e
c	c	b	e	a
b	b	e	a	c

*	e	a	c	b
e	e	a	c	b
a	a	c	b	e
c	c	b	e	a
b	b	e	a	c

Now we have a choice: each row and column needs an e and an a, but these could be arranged in two different ways while respecting inverses. Setting $b * b = a$ leads to the cyclic group again—can you work out how? Setting $b * b = e$ leads to the completed table below.

*	e	a	b	c
e	e	a	b	c
a	a	e	c	b
b	b	c	e	a
c	c	b	a	e

Is this a group? Yes: it is V from Section 6.8, the symmetry group of a non-square rectangle. And that's it. We have now explored all possibilities and established that *up to isomorphism*, there are exactly two groups of order four.

Next, groups of order three. Filling in a table's first row and column leaves only four blank cells. Check the possibilities for these cells and you will see that up to isomorphism, $\mathbb{Z}_3$ is the only group of order three.

*	e	a	b
e	e	a	b
a	a		
b	b		

*	e	a	b
e	e	a	b
a	a	b	e
b	b	e	a

It is even simpler to establish that $\mathbb{Z}_2$ is the only group of order two. Think about the table, and about where two-element groups have come up so far. Finally, there is one group with one element. Because a group must have an identity, the set will be $\{e\}$ and the operation $*$ will be defined by this table.

$$\begin{array}{c|c} * & e \\ \hline e & e \end{array}$$

We have seen this before via one-element subgroups. This structure satisfies closure because for every element in $\{e\}$ (all one of them), $e * e = e$. The identity is e, because for every element, $e * e = e * e = e$. The element e is its own inverse. And the operation is associative because $(e * e) * e = e * (e * e)$. As noted in Section 6.3, the idea of a one-element group might seem uninteresting or weird. If it seems uninteresting, fair enough. If it seems weird, remember not to let your understanding of the word *group* be contaminated by its everyday meaning. In mathematics we work with definitions, even if they apply in unexpected ways to degenerate cases.

To conclude this chapter, recall that Section 6.7 claimed that D_3 is the smallest non-commutative group. You now have enough information to work out why—can you put it together?

CHAPTER 7

Quotient Groups

This chapter introduces quotient groups, which arise when elements of a group 'clump together' to form sets that can themselves be treated as elements of another group. It observes where quotient groups arise and where they do not in cyclic groups and dihedral groups. It then formalizes the resulting ideas via cosets and normal subgroups, and links group, subgroup and quotient group orders via Lagrange's Theorem.

7.1 What is a quotient group?

Section 5.6 observed that D_3, the group of symmetries of an equilateral triangle, has an interesting structure. Its table splits naturally into four checkerboard squares: two 3×3 squares of rotations (with the identity) and two 3×3 squares of reflections.

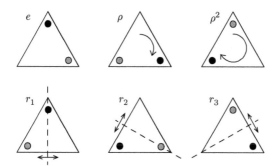

$\circ$	e	ρ	ρ^2	r_1	r_2	r_3
e	e	ρ	ρ^2	r_1	r_2	r_3
ρ	ρ	ρ^2	e	r_2	r_3	r_1
ρ^2	ρ^2	e	ρ	r_3	r_1	r_2
r_1	r_1	r_3	r_2	e	ρ^2	ρ
r_2	r_2	r_1	r_3	ρ	e	ρ^2
r_3	r_3	r_2	r_1	ρ^2	ρ	e

The same happens for other dihedral groups such as D_4, the group of symmetries of a square (see Section 6.8). If you constructed the table below when reading Section 5.6, you might want to get out your copy; if not, maybe photograph this one to refer to during this chapter.

$\circ$	e	ρ	ρ^2	ρ^3	r_1	r_2	r_3	r_4
e	e	ρ	ρ^2	ρ^3	r_1	r_2	r_3	r_4
ρ	ρ	ρ^2	ρ^3	e	r_3	r_4	r_2	r_1
ρ^2	ρ^2	ρ^3	e	ρ	r_2	r_1	r_4	r_3
ρ^3	ρ^3	e	ρ	ρ^2	r_4	r_3	r_1	r_2
r_1	r_1	r_4	r_2	r_3	e	ρ^2	ρ^3	ρ
r_2	r_2	r_3	r_1	r_4	ρ^2	e	ρ	ρ^3
r_3	r_3	r_1	r_4	r_2	ρ	ρ^3	e	ρ^2
r_4	r_4	r_2	r_3	r_1	ρ^3	ρ	ρ^2	e

In D_3 and D_4, the rotations block together because they form a subgroup (see Sections 6.2 and 6.8). But this phenomenon goes beyond that. The checkerboard pattern arises because composing a rotation with a reflection gives a reflection, and composing two reflections gives a rotation.

$\circ$	rotations	reflections
rotations	rotations	reflections
reflections	reflections	rotations

This works like a familiar structure: the odd and even numbers under addition. Adding two even numbers gives an even number; adding an even and an odd gives an odd; adding two odds gives an even. Formally, the even numbers form the subgroup $(2\mathbb{Z}, +)$ of the group $(\mathbb{Z}, +)$, and the evens and their complementary set of odds behave as in the middle table below. Moreover, the structures formed by the rotations and reflections in D_3 and by the evens and odds in $(\mathbb{Z}, +)$ are groups because both match the structure of the cyclic group $\mathbb{Z}_2$.

+	evens	odds
evens	evens	odds
odds	odds	evens

$+_2$	0	1
0	0	1
1	1	0

The 'elements' of these new groups are themselves sets: the rotations, for instance, form a single element. So this way of identifying a group 'within' a group is new. It is not like identifying subgroups because it is not elements of the original group that form another (usually smaller). Rather, *sets of elements* clump together to form 'elements' of the new one. When this happens, the subgroup and its complementary set are known as *cosets*. In D_3, the cosets of the subgroup $\{e, \rho, \rho^2\}$ are $\{e, \rho, \rho^2\}$ and $\{r_1, r_2, r_3\}$. In $\mathbb{Z}$, the cosets of the subgroup $2\mathbb{Z}$ are the even numbers and the odd numbers. The new type of group is called a *quotient group* or *factor group* because it is formed by 'dividing' the whole group by a subgroup. Using the language of division and the fact that structurally identical groups are *isomorphic*, we say that

$$D_3/\{e, \rho, \rho^2\} \cong \mathbb{Z}_2 \quad \text{('D_3 over $\{e, \rho, \rho^2\}$ is isomorphic to $\mathbb{Z}_2$');}$$
$$\mathbb{Z}/2\mathbb{Z} \cong \mathbb{Z}_2 \quad \text{('$\mathbb{Z}$ over $2\mathbb{Z}$ is isomorphic to $\mathbb{Z}_2$').}$$

Note that in the two quotient groups $D_3/\{e, \rho, \rho^2\}$ and $\mathbb{Z}/2\mathbb{Z}$, the subgroup plays the role of the identity. But the quotient group operations merit some thought because they involve not composing individual symmetries or adding individual numbers but operating on sets. That causes no problems here, but the general case is more complex and will be addressed in Section 7.6.

For now, the obvious question is, does every group split naturally into a subgroup and other cosets that together form a quotient group? Or is there something special about rotations in dihedral groups and the even numbers in $\mathbb{Z}$? Are there, for instance, quotient groups not isomorphic to $\mathbb{Z}_2$, or do quotient groups arise only when the subgroup is 'half of' the group? What do you think?

7.2 Quotient groups in cyclic groups

You might have observed that the quotient group $\mathbb{Z}/2\mathbb{Z}$ formed by the even and odd numbers *is* $\mathbb{Z}_2$, because elements of $\mathbb{Z}_2$ are congruence classes of the integers modulo 2 under addition modulo 2 (see Sections 3.3 and 3.4). In the table for $\mathbb{Z}_2$, the number 0 represents the set $\{\ldots, -6, -4, -2, 0, 2, 4, 6, \ldots\}$ (the evens), the number 1 represents the set $\{\ldots, -5, -3, -1, 1, 3, 5, 7, \ldots\}$ (the odds), and addition captures remainders on division by 2.

+	evens	odds
evens	even	odds
odds	odds	evens

$+_2$	0	1
0	0	1
1	1	0

This interpretation raises a question: if $\mathbb{Z}/2\mathbb{Z} \cong \mathbb{Z}_2$, is it also true that $\mathbb{Z}/3\mathbb{Z} \cong \mathbb{Z}_3$, that $\mathbb{Z}/4\mathbb{Z} \cong \mathbb{Z}_4$, and in general that $\mathbb{Z}/n\mathbb{Z} \cong \mathbb{Z}_n$?

The answer is yes, which is nice because $\mathbb{Z}/n\mathbb{Z} \cong \mathbb{Z}_n$ looks tidy. But we should understand this isomorphism rather than just wanting to believe in it. For instance, consider $(3\mathbb{Z}, +)$, the subgroup of $(\mathbb{Z}, +)$ comprising all multiples of 3. All elements of this subgroup leave remainder 0 on division by 3. Under the congruence class interpretation, the group $(\mathbb{Z}, +)$ splits into this subgroup and another *two* cosets: numbers leaving remainder 1 on division by 3 and numbers leaving remainder 2. These three cosets form a quotient group isomorphic to $\mathbb{Z}_3$.

+	remainder 0	remainder 1	remainder 2
remainder 0	remainder 0	remainder 1	remainder 2
remainder 1	remainder 1	remainder 2	remainder 0
remainder 2	remainder 2	remainder 0	remainder 1

$+_3$	0	1	2
0	0	1	2
1	1	2	0
2	2	0	1

These ideas can be formalized using the definition of *coset*, first mentioned in Section 2.6. Here it is, in multiplicative and additive formulations.

Definition: Suppose that H is a subgroup of G. Then the **left coset** of H containing a is $aH = \{ah|h \in H\}$.

Definition: Suppose that H is a subgroup of G. Then the **left coset** of H containing a is $a + H = \{a + h|h \in H\}$.

In $G = \mathbb{Z}$, the additive formulation is more natural. And applying the definition to the subgroup $H = 3\mathbb{Z}$ gives exactly the expected cosets.

$$0 + 3\mathbb{Z} = \{0 + z|z \in 3\mathbb{Z}\} = \{\ldots, -6, -3, 0, 3, 6, 9, \ldots\};$$
$$1 + 3\mathbb{Z} = \{1 + z|z \in 3\mathbb{Z}\} = \{\ldots, -5, -2, 1, 4, 7, 10, \ldots\};$$
$$2 + 3\mathbb{Z} = \{2 + z|z \in 3\mathbb{Z}\} = \{\ldots, -4, -1, 2, 5, 8, 11, \ldots\}.$$

If you wondered why the definitions say *left coset* instead of just *coset*, good. Try working out what the *right cosets* would be—are they different? Then observe that $0 + 3\mathbb{Z} = 3\mathbb{Z}$, and that denoting the cosets by $3\mathbb{Z}$, $1 + 3\mathbb{Z}$ and $2 + 3\mathbb{Z}$ captures the fact that they are sets, in a brief way that is useful in tables.

'+'	$3\mathbb{Z}$	$1 + 3\mathbb{Z}$	$2 + 3\mathbb{Z}$
$3\mathbb{Z}$	$3\mathbb{Z}$	$1 + 3\mathbb{Z}$	$2 + 3\mathbb{Z}$
$1 + 3\mathbb{Z}$	$1 + 3\mathbb{Z}$	$2 + 3\mathbb{Z}$	$3\mathbb{Z}$
$2 + 3\mathbb{Z}$	$2 + 3\mathbb{Z}$	$3\mathbb{Z}$	$1 + 3\mathbb{Z}$

Do you now believe that $\mathbb{Z}/n\mathbb{Z} \cong \mathbb{Z}_n$ for every natural number n? If not, think through cosets and tables for more values of n. Once you are convinced, you know about every quotient group in $(\mathbb{Z}, +)$, because every subgroup of $(\mathbb{Z}, +)$ takes the form $(n\mathbb{Z}, +)$ (see Section 6.4). And this reasoning addresses Section 6.5's question about the relationship between $n\mathbb{Z}$ and $\mathbb{Z}_n$. The group $n\mathbb{Z}$ is an infinite group under standard addition; the group $\mathbb{Z}_n$ is a finite group with n elements under addition modulo n. These are different, but we have now established their important relationship.

Indeed, we can extend this reasoning to finite cyclic groups. For instance, $\mathbb{Z}_{12}$ has a subgroup isomorphic to $\mathbb{Z}_4$—see Section 6.2. What structure do you think $\mathbb{Z}_{12}/\mathbb{Z}_4$ has? Feels like it should be $\mathbb{Z}_3$, no? Fortunately, this is true. To see why, recall that the subgroup of $\mathbb{Z}_{12}$ isomorphic to $\mathbb{Z}_4$ is $H = (\{0,3,6,9\}, +_{12})$.

$+_{12}$	0	1	2	3	4	5	6	7	8	9	10	11
0	0	1	2	3	4	5	6	7	8	9	10	11
1	1	2	3	4	5	6	7	8	9	10	11	0
2	2	3	4	5	6	7	8	9	10	11	0	1
3	3	4	5	6	7	8	9	10	11	0	1	2
4	4	5	6	7	8	9	10	11	0	1	2	3
5	5	6	7	8	9	10	11	0	1	2	3	4
6	6	7	8	9	10	11	0	1	2	3	4	5
7	7	8	9	10	11	0	1	2	3	4	5	6
8	8	9	10	11	0	1	2	3	4	5	6	7
9	9	10	11	0	1	2	3	4	5	6	7	8
10	10	11	0	1	2	3	4	5	6	7	8	9
11	11	0	1	2	3	4	5	6	7	8	9	10

The cosets of H in $\mathbb{Z}_{12}$ can be calculated using the definition.

$$0 + H = \{0 + h \mid h \in H\} = \{0,3,6,9\};$$
$$1 + H = \{1 + h \mid h \in H\} = \{1,4,7,10\};$$
$$2 + H = \{2 + h \mid h \in H\} = \{2,5,8,11\}.$$

But these cosets do not show up well in the table above. Quotient group structures were evident in Section 7.1 because the cosets were in 'blocks'. Inspired by that, we can rearrange as below; check that the individual additions are correct.

$+_{12}$	0	3	6	9	1	4	7	10	2	5	8	11
0	0	3	6	9	1	4	7	10	2	5	8	11
3	3	6	9	0	4	7	10	1	5	8	11	2
6	6	9	0	3	7	10	1	4	8	11	2	5
9	9	0	3	6	10	1	4	7	11	2	5	8
1	1	4	7	10	2	5	8	11	3	6	9	0
4	4	7	10	1	5	8	11	2	6	9	0	3
7	7	10	1	4	8	11	2	5	9	0	3	6
10	10	1	4	7	11	2	5	8	0	3	6	9
2	2	5	8	11	3	6	9	0	4	7	10	1
5	5	8	11	2	6	9	0	3	7	10	1	4
8	8	11	2	5	9	0	6	3	10	1	4	7
11	11	2	5	8	0	3	6	9	1	4	7	10

Then take a moment to be impressed by the structure. Adding any element from the coset $\{0,3,6,9\}$ to any from the coset $\{1,4,7,10\}$ gives one from the coset $\{1,4,7,10\}$. And adding any element from the coset $\{0,3,6,9\}$ to any from the coset $\{2,5,8,11\}$ gives one from the coset $\{2,5,8,11\}$. That is what it means for the subgroup $\{0,3,6,9\}$ to be the identity in the quotient group. Moreover, adding any element from the coset $\{1,4,7,10\}$ to any from the coset $\{2,5,8,11\}$ gives one from the coset $\{0,3,6,9\}$, and so on. Formally, the operation on cosets is *well defined*, so we can write the table in more condensed ways.

$+_{12}$	$\{0,3,6,9\}$	$\{1,4,7,10\}$	$\{2,5,8,11\}$
$\{0,3,6,9\}$	$\{0,3,6,9\}$	$\{1,4,7,10\}$	$\{2,5,8,11\}$
$\{1,4,7,10\}$	$\{1,4,7,10\}$	$\{2,5,8,11\}$	$\{0,3,6,9\}$
$\{2,5,8,11\}$	$\{2,5,8,11\}$	$\{0,3,6,9\}$	$\{1,4,7,10\}$

$+_{12}$	H	$1+H$	$2+H$
H	H	$1+H$	$2+H$
$1+H$	$1+H$	$2+H$	H
$2+H$	$2+H$	H	$1+H$

With that in place, can you generalize? What are some other subgroups of $\mathbb{Z}_{12}$? What would be their quotient groups in $\mathbb{Z}_{12}$? How would this work in other cyclic groups, like $\mathbb{Z}_{60}$, or $\mathbb{Z}_7$? The latter has only two subgroups: the whole group ($\mathbb{Z}_7, +_7$) and the trivial subgroup ($\{0\}, +_7$) (see Section 6.3). What are their quotient groups in $\mathbb{Z}_7$?

To conclude this section, a comment on learning about quotient groups and cosets. When coset definitions are introduced, you will likely be asked to calculate cosets for specific subgroups of specific groups. This is not very interesting as it just involves calculating lists. It is worth doing, to develop your sense of how cosets are formed, and to understand that in aH or $a+H$, the element a can be any element of the group, not only—as muddled students sometimes think—elements of the subgroup H (what happens if $a \in H$?). But do not make the mistake of thinking page-filling calculations important. What is important and interesting about cosets is their role as elements of quotient groups. By all means calculate attentively, but then make some effort to think of the resulting cosets as single entities. This compression is important.

7.3 Element–coset commutativity

If you understand cyclic groups, the previous section might leave you unimpressed. With addition based on division with remainders, perhaps it is obvious that quotient groups work tidily. But that only makes the general question more interesting. Quotient groups arise not only in cyclic groups but also in dihedral groups, where the elements are not numbers but symmetries. Symmetries have neither division nor remainders, at least in the usual sense—could there be an analogue of division with remainders for dihedral groups? And, to reiterate the earlier question, does a quotient group arise for every subgroup of every group, or not?

We will approach these questions by considering the dihedral group D_4. The table in Section 7.1 (refer to your copy) shows that if $H = \{e, \rho, \rho^2, \rho^3\}$

is the four-element subgroup of rotations, then $D_4/H \cong \mathbb{Z}_2$. But D_4 has many subgroups, as discussed in Section 6.6 and represented below. Do other subgroups give rise to meaningful cosets and a quotient group? Does $H = \{e, \rho^2\}$ give rise to a quotient group in D_4, for instance? If so, how many elements would you expect that quotient group to have?

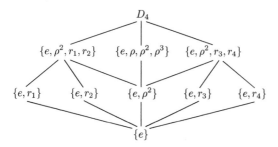

Calculating cosets is straightforward, but we should think carefully about those calculations because for cyclic groups I implicitly relied on experience of division with remainders. For instance, I calculated the three cosets of $3\mathbb{Z}$ shown below, then stopped. Probably three seemed like the right number. But why, exactly? What would happen if we calculated $3 + 3\mathbb{Z}, 4 + 3\mathbb{Z}$, and so on?

$$0 + 3\mathbb{Z} = \{0 + z | z \in 3\mathbb{Z}\} = \{\ldots, -6, -3, 0, 3, 6, 9, \ldots\};$$
$$1 + 3\mathbb{Z} = \{1 + z | z \in 3\mathbb{Z}\} = \{\ldots, -5, -2, 1, 4, 7, 10, \ldots\};$$
$$2 + 3\mathbb{Z} = \{2 + z | z \in 3\mathbb{Z}\} = \{\ldots, -4, -1, 2, 5, 8, 11, \ldots\}.$$

For dihedral groups, most people have little experience. So, although your sense of how elements 'clump together' might lead you to answer that $H = \{e, \rho^2\}$ should have four two-element cosets in the eight-element group D_4, it is less likely that you can intuit what those cosets should be. For that reason, we will use the definition to calculate cosets for all eight elements, then examine the results. Here is the definition again—this time, the multiplicative version feels more natural.

Definition: Suppose that H is a subgroup of G. Then the **left coset** of H containing a is $aH = \{ah | h \in H\}$.

The calculations appear below (I filled these in using the table).

$$eH = \{ee,\ e\rho^2\} \quad = \{e, \rho^2\};$$
$$\rho H = \{\rho e,\ \rho\rho^2\} \quad = \{\rho, \rho^3\};$$
$$\rho^2 H = \{\rho^2 e,\ \rho^2\rho^2\} = \{\rho^2, e\};$$
$$\rho^3 H = \{\rho^3 e,\ \rho^3\rho^2\} = \{\rho^3, \rho\};$$
$$r_1 H = \{r_1 e,\ r_1\rho^2\} \quad = \{r_1, r_2\};$$
$$r_2 H = \{r_2 e,\ r_2\rho^2\} \quad = \{r_2, r_1\};$$
$$r_3 H = \{r_3 e,\ r_3\rho^2\} \quad = \{r_3, r_4\};$$
$$r_4 H = \{r_4 e,\ r_4\rho^2\} \quad = \{r_4, r_3\}.$$

This gives four distinct cosets, each appearing twice (check). But notice that it could have gone wrong. If the cosets overlapped—if, say, we had found that $r_1 H = \{r_1, r_2\}$ but $r_2 H = \{r_2, r_3\}$—then the cosets would not *partition* the group (see Section 3.4): D_4 would not split into sets that could potentially form elements of a quotient group. However, everything looks okay, and to visualize a potential quotient group structure we can rearrange the elements in a table for D_4. First, we can put the subgroup $H = \{e, \rho^2\}$ in a 'block'. This block is self-contained because a subgroup must be closed under the group operation.

$\circ$	e	ρ^2
e	e	ρ^2
ρ^2	ρ^2	e

The next listed coset is $\{\rho, \rho^3\}$, so that can go in a second block. Check that the cells below are filled in correctly, then attend to commutativity. This part of the table is symmetric across the main diagonal because all the rotations commute. That might seem unremarkable, but it will be important in a minute.

$\circ$	e	ρ^2	ρ	ρ^3
e	e	ρ^2	ρ	ρ^3
ρ^2	ρ^2	e	ρ^3	ρ
ρ	ρ	ρ^3	ρ^2	e
ρ^3	ρ^3	ρ	e	ρ^2

The next listed coset is $\{r_1, r_2\}$, which can go in a third block. Completing more cells maintains commutativity, because the identity commutes with everything, and so does the 180° rotation ρ^2 (the subgroup $\{e, \rho^2\}$ is the *centre* of D_4—see Section 6.10).

∘	e	ρ^2	ρ	ρ^3	r_1	r_2
e	e	ρ^2	ρ	ρ^3	r_1	r_2
ρ^2	ρ^2	e	ρ^3	ρ	r_2	r_1
ρ	ρ	ρ^3	ρ^2	e		
ρ^3	ρ^3	ρ	e	ρ^2		
r_1	r_1	r_2				
r_2	r_2	r_1				

After that, it gets interesting. Completing more cells shows that commutativity continues to matter, but no longer at the level of single elements. For instance, it is not true that $r_1 \rho = \rho r_1$ (check). But it *is* true that

$$r_1\{\rho, \rho^3\} = \{r_3, r_4\} = \{\rho, \rho^3\}r_1 \quad \text{and} \quad r_2\{\rho, \rho^3\} = \{r_3, r_4\} = \{\rho, \rho^3\}r_2.$$

So, although the table for individual elements is not symmetrical across the main diagonal, it nevertheless splits into intact coset blocks. All four of $r_1\{\rho, \rho^3\}$, $\{\rho, \rho^3\}r_1$, $r_2\{\rho, \rho^3\}$ and $\{\rho, \rho^3\}r_2$ give the same coset, $\{r_3, r_4\}$.

∘	e	ρ^2	ρ	ρ^3	r_1	r_2
e	e	ρ^2	ρ	ρ^3	r_1	r_2
ρ^2	ρ^2	e	ρ^3	ρ	r_2	r_1
ρ	ρ	ρ^3	ρ^2	e	r_3	r_4
ρ^3	ρ^3	ρ	e	ρ^2	r_4	r_3
r_1	r_1	r_2	r_4	r_3		
r_2	r_2	r_1	r_3	r_4		

And the same thing happens throughout the table—individual elements 'commute' with cosets and the cosets stay in blocks.

∘	e	ρ^2	ρ	ρ^3	r_1	r_2	r_3	r_4
e	e	ρ^2	ρ	ρ^3	r_1	r_2	r_3	r_4
ρ^2	ρ^2	e	ρ^3	ρ	r_2	r_1	r_4	r_3
ρ	ρ	ρ^3	ρ^2	e	r_3	r_4	r_2	r_1
ρ^3	ρ^3	ρ	e	ρ^2	r_4	r_3	r_1	r_2
r_1	r_1	r_2	r_4	r_3	e	ρ^2	ρ^3	ρ
r_2	r_2	r_1	r_3	r_4	ρ^2	e	ρ	ρ^3
r_3	r_3	r_4	r_1	r_2	ρ	ρ^3	e	ρ^2
r_4	r_4	r_3	r_2	r_1	ρ^3	ρ	ρ^2	e

Thus D_4 splits into a potential quotient group for the subgroup $H = \{e, \rho^2\}$. This potential quotient group has four elements, each a two-element coset of $\{e, \rho^2\}$.

∘	$\{e, \rho^2\}$	$\{\rho, \rho^3\}$	$\{r_1, r_2\}$	$\{r_3, r_4\}$
$\{e, \rho^2\}$	$\{e, \rho^2\}$	$\{\rho, \rho^3\}$	$\{r_1, r_2\}$	$\{r_3, r_4\}$
$\{\rho, \rho^3\}$	$\{\rho, \rho^3\}$	$\{e, \rho^2\}$	$\{r_3, r_4\}$	$\{r_1, r_2\}$
$\{r_1, r_2\}$	$\{r_1, r_2\}$	$\{r_3, r_4\}$	$\{e, \rho^2\}$	$\{\rho, \rho^3\}$
$\{r_3, r_4\}$	$\{r_3, r_4\}$	$\{r_1, r_2\}$	$\{\rho, \rho^3\}$	$\{e, \rho^2\}$

∘	H	ρH	$r_1 H$	$r_3 H$
H	H	ρH	$r_1 H$	$r_3 H$
ρH	ρH	H	$r_3 H$	$r_1 H$
$r_1 H$	$r_1 H$	$r_3 H$	H	ρH
$r_3 H$	$r_3 H$	$r_1 H$	ρH	H

Now, why do you think I refer to this as a 'potential' quotient group? With cosets as elements, everything seems to work. But there are ways to fill in tables without forming groups (see Section 6.11). Does this way satisfy the group axioms? Yes: it must, because it is isomorphic to the Klein four-group V (see Section 6.8). Thus we can write $D_4/\{e, \rho^2\} \cong V$.

7.4 Left and right cosets

Perhaps, then, a quotient group arises for every subgroup of every group. But we should not conclude too hastily. Everything worked for $H = \{e, \rho^2\}$ in D_4 due to an element–coset form of commutativity. But, with respect to commutativity, some elements are less 'well behaved' than ρ^2. So it is worth checking similarly for different subgroups. Calculating cosets for $H = \{e, r_1\}$, for instance, gives:

$$eH = \{ee,\ er_1\} \quad = \{e, r_1\};$$
$$\rho H = \{\rho e,\ \rho r_1\} \quad = \{\rho, r_3\};$$
$$\rho^2 H = \{\rho^2 e,\ \rho^2 r_1\} = \{\rho^2, r_2\};$$
$$\rho^3 H = \{\rho^3 e,\ \rho^3 r_1\} = \{\rho^3, r_4\};$$
$$r_1 H = \{r_1 e,\ r_1 r_1\} \quad = \{r_1, e\};$$
$$r_2 H = \{r_2 e,\ r_2 r_1\} \quad = \{r_2, \rho^2\};$$
$$r_3 H = \{r_3 e,\ r_3 r_1\} \quad = \{r_3, \rho\};$$
$$r_4 H = \{r_4 e,\ r_4 r_1\} \quad = \{r_4, \rho^3\}.$$

Again this yields four distinct cosets, each appearing twice. And again $H = \{e, r_1\}$ is a subgroup of D_4, so it is closed under composition and forms a first table block. This time, the next listed coset, $\{\rho, r_3\}$, gives a self-contained block below the subgroup—check that the partial table below is filled in correctly. However, to the right of the subgroup, things 'go wrong'. The top row reflects the fact that $e\{\rho, r_3\} = \{\rho, r_3\}$. But the lower row contains $r_1 \rho = r_4$, not r_3, and $r_1 r_3 = \rho^3$, not ρ. In other words, $\rho\{e, r_1\} \neq \{e, r_1\}\rho$.

$\circ$	e	r_1	ρ	r_3
e	e	r_1	ρ	r_3
r_1	r_1	e	r_4	ρ^3
ρ	ρ	r_3		
r_3	r_3	ρ		

This establishes that not all elements of D_4 'commute' with cosets of $\{e, r_1\}$, and that the cosets do not form self-contained blocks acting as elements in a quotient group. Thus there *is* something special about some subgroups in some groups. In D_4, the subgroups $\{e, \rho, \rho^2, \rho^3\}$ and $\{e, \rho^2\}$ give rise to quotient groups, but the subgroup $\{e, r_1\}$ does not. Subgroups that give rise to quotient groups thus merit a name, and we call them *normal subgroups*. I do not know the history but I find this an appealingly snooty use of language, implying that it is *normal* for a subgroup to give rise to a quotient group and that subgroups that do not are inferior. They are not inferior, of course—any set that meets the definition of a subgroup is a subgroup. But normal subgroups do have extra properties. To understand those properties, we will work out what distinguishes normal subgroups from other subgroups.

For all subgroups, calculating cosets is straightforward from the definition. Moreover, in the cases considered in this chapter, cosets *partition* their group: they are disjoint and leave nothing out. This is necessary for a normal subgroup because if cosets do not partition a group, they cannot form self-contained elements in a quotient group. Must partitioning therefore be part of a definition of normal subgroup? Or does it work for all subgroups anyway? What do you think?

For me the answer is not intuitively obvious—groups can differ greatly, so the fact that cosets of some subgroups partition their groups does not obviously generalize. However, it turns out that the cosets of any subgroup do partition the group, which can be proved by generalizing an argument from Section 3.4. First, (left) cosets leave nothing out because for every group G, every subgroup H contains the identity e. So, for every element $a \in G$, the element $a = ae$ is in the coset aH. One way to prove that cosets are disjoint is to prove that for any two cosets aH and bH, either $aH \cap bH = \emptyset$ (they don't overlap at all) or $aH = bH$ (they overlap completely, so they are the same coset).

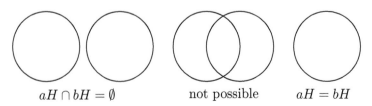

$aH \cap bH = \emptyset$ not possible $aH = bH$

As in Section 3.4, the 'either ... or' statement can be proved by demonstrating that if $aH \cap bH \neq \emptyset$, then $aH = bH$, and a general theorem and proof appears below. If the abstraction makes it hard going, reread Section 3.4 first, noting that the operation in $12\mathbb{Z}$ is addition, so cosets take the form $a + 12\mathbb{Z}$, whereas here the argument is general, so the operation is denoted by juxtaposition and cosets take the form aH.

Theorem: Suppose that H is a subgroup of G.

Then the left cosets of H in G partition G.

Proof: Let $a \in G$.

Observe that $e \in H$.

So $a = ae \in aH$.

Thus every element appears in a coset.

Now suppose that $b \in G$ and $aH \cap bH \neq \emptyset$.

Then $\exists x \in aH \cap bH$.

Because $x \in aH$, $\exists h_1 \in H$ such that $x = ah_1$.

Because $x \in bH$, $\exists h_2 \in H$ such that $x = bh_2$.

So $ah_1 = bh_2$ and, because h_1 and h_2 have inverses (in H), $b = ah_1 h_2^{-1}$ and $a = bh_2 h_1^{-1}$.

Now let $y \in bH$.

Then $\exists h_3 \in H$ such that $y = bh_3 = ah_1 h_2^{-1} h_3$.

Thus $y \in aH$ because $h_1 h_2^{-1} h_3 \in H$ by subgroup closure.

So $bH \subseteq aH$.

Similarly let $z \in aH$ be arbitrary.

Then $\exists h_3 \in H$ such that $z = ah_3 = bh_2 h_1^{-1} h_3$.

Thus $z \in bH$ because $h_1 h_2^{-1} h_3 \in H$ by subgroup closure.

So $aH \subseteq bH$.

Hence $bH \subseteq aH$ and $aH \subseteq bH$, so $aH = bH$.

Thus we have proved that if $aH \cap bH \neq \emptyset$ then $aH = bH$.

Hence the cosets of H in G partition G.

Perhaps you can work out how to adjust the above proof to work for right cosets, where these are defined as below. I recommend trying that.

Definition: Suppose that H is a subgroup of G. Then the **right coset** of H containing a is $Ha = \{ha | h \in H\}$.

Thus both left cosets and right cosets partition their groups. And this is true for all subgroups, not just normal ones. What makes normal subgroups special—as you might anticipate—is how left and right cosets work together. This can be clarified by examining right cosets of $\{e, r_1\}$ as a subgroup of D_4. The definition above yields four distinct right cosets that partition the group.

$$
\begin{aligned}
He &= \{ee,\ r_1 e\} &&= \{e, r_1\}; \\
H\rho &= \{e\rho,\ r_1\rho\} &&= \{\rho, r_4\}; \\
H\rho^2 &= \{e\rho^2,\ r_1\rho^2\} &&= \{\rho^2, r_2\}; \\
H\rho^3 &= \{e\rho^3,\ r_1\rho^3\} &&= \{\rho^3, r_3\}; \\
Hr_1 &= \{er_1, r_1 r_1\} &&= \{r_1, e\}; \\
Hr_2 &= \{er_2, r_1 r_2\} &&= \{r_2, \rho^2\}; \\
Hr_3 &= \{er_3, r_1 r_3\} &&= \{r_3, \rho^3\}; \\
Hr_4 &= \{er_4, r_1 r_4\} &&= \{r_4, \rho\}.
\end{aligned}
$$

But the left and right cosets are *not the same.*

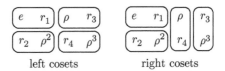

left cosets   right cosets

This led to the problem in constructing a quotient group table for $\{e, r_1\}$ in D_4—look back to check. In general, if left and right cosets are unequal then constructing a quotient group 'goes wrong'. This motivates the definition of normal subgroup.

7.5 Normal subgroups: theory

We have now established that a subgroup is *normal*—that its cosets act as elements of a quotient group—only if the left and right cosets 'match'. This is captured in the definition of *normal subgroup*.[1]

Definition: Let H be a subgroup of G. Then H is a **normal subgroup** if and only if $\forall a \in G, aH = Ha$.

Note the quantifier: every element of the group must 'commute' with the subgroup. Also, normality is a property of subgroups in relation to groups. It makes no sense to speak of 'normal groups', and we should state that a subgroup is normal only if it is clear what 'main' group is under discussion. Mathematicians will forgive repetition: 'Recall that H is a normal subgroup of G' or 'Because $\{e, \rho^2\}$ is a normal subgroup of $D_4, \ldots$'. They will not readily forgive ambiguity: 'H is a normal subgroup' might make mathematicians think 'Of what?'

Normal subgroups, being special, have their own notation: 'H is a normal subgroup of G' can be written $H \trianglelefteq G$, and if H is both normal and proper we can write $H \triangleleft G$. A normal subgroup might be denoted by N, so you might see—or want to write for your own notes—an abbreviated version of the definition.

Definition: $N \trianglelefteq G$ iff $\forall a \in G, aN = Na$.

Indeed, you might see various non-abbreviated versions; alternative formulations are discussed later in this section. First, however, a point of logic. Section 7.4 established that normality of H in G is *necessary* for G/H to be a quotient group—no normality, no quotient group. We have not explored whether normality is also *sufficient*.

$$\text{quotient group } G/H \Rightarrow H \text{ normal}$$
$$\text{quotient group } G/H \underset{?}{\Leftarrow} H \text{ normal}$$

---

[1] If you have opened this book here because you are stuck with normal subgroups, I recommend reading this chapter from the beginning.

Moreover, you might have noticed that the equation $aH = Ha$ relates only to a single row and column in a potential quotient group table, which is not quite the same as cosets 'hanging together' in blocks. Happily, everything does work: normality is both necessary and sufficient for a subgroup to give rise to a quotient group. This can be established in part by generalizing a theorem and proof from Section 3.3, which showed that addition for cosets of $12\mathbb{Z}$ in $\mathbb{Z}$ is *well defined*—that the natural operation on cosets respects the results of operating on elements. The general version below shows that the natural operation on cosets of a normal subgroup is always well defined—that, for cosets of a normal subgroup (written in multiplicative notation), it is always meaningful to write $(aH)(bH) = (ab)H$. As you read the proof, notice that $bH = Hb$ does not imply that $h_1 b = bh_1$, which might be a tempting inference. Because H is a set, it implies only that there is a possibly different element $h_3 \in H$ such that $h_1 b = bh_3$. Where would the proof break down if a subgroup were not normal?

Theorem: Suppose that H is a normal subgroup of G and that $x \in aH$ and $y \in bH$. Then $xy \in (ab)H$.

Proof: Suppose that H is a normal subgroup of G.

Suppose that $x \in aH$ so $\exists h_1 \in H$ such that $x = ah_1$

and $\quad\quad y \in bH$ so $\exists h_2 \in H$ such that $y = bh_2$.

Then $xy = (ah_1)(bh_2) = a(h_1 b)h_2$ by associativity.

Now, because H is normal, we know that $bH = Hb$.

So $\exists h_3 \in H$ such that $h_1 b = bh_3$.

Hence $xy = a(bh_3)h_2 = (ab)(h_3 h_2)$, again by associativity.

But $h_3 h_2 \in H$ because H is closed under the group operation.

Thus $xy \in (ab)H$.

The fact that the coset operation is well defined is crucial for a quotient group to exist—it means that, under the natural operation, the cosets act as meaningful single elements. What else is needed? A quotient group

G/H—like any group—must satisfy the group definition. So it must be closed under the operation, which it is because for any two cosets aH and bH, $(aH)(bH) = (ab)H$ (another coset). Associativity requires that for any three cosets aH, bH and cH, $((aH)(bH))cH = aH((bH)(cH))$. How would you show that this equality holds? The identity coset will be H or, if you prefer, eH. Why is eH equal to H, and why is it true that for every coset aH, $(eH)(aH) = (aH)(eH) = aH$? Finally, which coset do you think will act as the inverse of aH? This is not a trick question—you can check that the obvious choice works. In your course, look out for a formally written proof that captures the ideas in this paragraph; this establishes that normality of H in G is sufficient for G/H to be a quotient group.

Now, I mentioned earlier that the normal subgroup definition can be formulated in several ways. I like the $aH = Ha$ formulation—repeated below—because for me it links to the intuitive understanding built up in this chapter.

Definition: Let H be a subgroup of G. Then H is a **normal subgroup** if and only if $\forall a \in G$, $aH = Ha$.

But your course might use a different formulation, where three are listed below. Think about object types: where does each use single elements and sets? Can you see that they are all logically equivalent?

Definition: Let H be a subgroup of G. Then H is a **normal subgroup** if and only if $\forall a \in G$, $\forall h \in H$, $\exists h' \in H$ such that $ah = h'a$.

Definition: Let H be a subgroup of G. Then H is a **normal subgroup** if and only if $\forall a \in G$, $a^{-1}Ha = H$.

Definition: Let H be a subgroup of G. Then H is a **normal subgroup** if and only if $\forall a \in G$, $\forall h \in H$, $a^{-1}ha \in H$.

Equivalence might be proved in your course, and means that it does not matter which formulation is treated as the definition and which are proved as theorems. But obviously these are easily muddled, so keep your objects and quantifiers clear.

It might also be useful to know that coset definitions too can be formulated differently. I defined cosets as below.

Definition: Suppose that H is a subgroup of G. Then the **left coset** of H containing a is $aH = \{ah|h \in H\}$.

Definition: Suppose that H is a subgroup of G. Then the **left coset** of H containing a is $a + H = \{a + h|h \in H\}$.

But saying that $x \in a + H$ is equivalent to saying that there exists $h \in H$ such that $x = a + h$, or to saying that $x - a \in H$. Either draws attention to the fact that elements within cosets differ by elements of the subgroup. I recommend thinking about how these formulations relate to the diagram below (from Section 3.4) showing vertical representations for the cosets of $H = 3\mathbb{Z}$ in $G = \mathbb{Z}$. Each column is a coset; the leftmost is the subgroup $3\mathbb{Z}$. Within each coset, elements differ from one another by elements of the subgroup. And each coset is horizontally 'offset' from the subgroup by a fixed number—the remainder on division by 3.

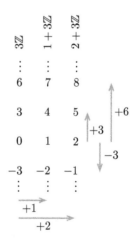

7.6 Normal subgroups: examples

The previous section was mostly formal, and your course might be like that—Abstract Algebra can be presented primarily in terms of definitions, theorems and proofs. But I like to feel that mathematics is *about* something—I like examples front and centre. So we will explore some now, viewing them through the lens of the theory.

First, in abelian groups, all subgroups are normal. That is because groups in which the operation is commutative automatically have element–coset commutativity. An abelian group G has $ab = ba$ for every $a, b \in G$. So, if it has a subgroup H, it must be true that $ah = ha$ for every $a \in G$ and $h \in H$, and thus that $aH = Ha$ for every $a \in G$. For instance, every subgroup of every cyclic group is normal, and there is a chain of normal subgroups $(\mathbb{Z}, +) \lhd (\mathbb{Q}, +) \lhd (\mathbb{R}, +) \lhd (\mathbb{C}, +)$.

Second, some groups are not abelian but do have 'commutative bits'. The *centre* of a group comprises the elements that commute with everything.

Definition: The **centre** of a group G is the set $\{x \in G | \forall g \in G, xg = gx\}$.

The centre is always a subgroup—see Section 6.10. It is always normal because if H is the centre of G, then $ah = ha$ for every $a \in G$ and $h \in H$, and thus again $aH = Ha$ for every $a \in G$. This is one way to identify normal subgroups in non-abelian groups. For instance, the subgroup $\{e, \rho^2\}$ is the centre of D_4 and is, as established directly in Section 7.3, a normal subgroup.

Third, the definition of normal subgroup requires only that $aH = Ha$ for every $a \in G$, not that $ah = ha$ for every $a \in G$ and $h \in H$. Can you see the difference? As in the previous section, $aH = Ha$ means only that there exist $h_1, h_2 \in H$ such that $ah_1 = h_2a$; the subgroup elements h_1 and h_2 can differ. Where have we seen a normal subgroup that is bigger than the centre of its group? One appeared in Section 7.1. For the subgroup $H = \{e, \rho, \rho^2, \rho^3\}$ of D_4, it is not true that $ah = ha$ for every $a \in G$ and $h \in H$. For instance, $r_1\rho \neq \rho r_1$. But it *is* true that $aH = Ha$ for every $a \in G$. For instance, $r_1\{e, \rho, \rho^2, \rho^3\} = \{r_1, r_2, r_3, r_4\} = \{e, \rho, \rho^2, \rho^3\}r_1$. This led to the quotient group structure $D_4/\{e, \rho, \rho^2, \rho^3\} \cong \mathbb{Z}_2$.

∘	e	ρ	ρ^2	ρ^3	r_1	r_2	r_3	r_4
e	e	ρ	ρ^2	ρ^3	r_1	r_2	r_3	r_4
ρ	ρ	ρ^2	ρ^3	e	r_3	r_4	r_2	r_1
ρ^2	ρ^2	ρ^3	e	ρ	r_2	r_1	r_4	r_3
ρ^3	ρ^3	e	ρ	ρ^2	r_4	r_3	r_1	r_2
r_1	r_1	r_4	r_2	r_3	e	ρ^2	ρ^3	ρ
r_2	r_2	r_3	r_1	r_4	ρ^2	e	ρ	ρ^3
r_3	r_3	r_1	r_4	r_2	ρ	ρ^3	e	ρ^2
r_4	r_4	r_2	r_3	r_1	ρ^3	ρ	ρ^2	e

Fourth, this table can help us to see that if a subgroup's order is half of its group's (finite) order, then the subgroup must be normal. Can you see why? Remember that in a group table, every element appears once in each row and column (see Section 5.3). So, if the top left quadrant of the table contains subgroup elements, the top right and bottom left quadrants must contain non-subgroup elements. What does that leave for the bottom right quadrant? The subgroup elements again. Thus the table must be arranged in a 2 × 2 checkerboard pattern of subgroup and non-subgroup elements, giving a quotient group isomorphic to $\mathbb{Z}_2$. In $\mathbb{Z}_{12}$, for instance, the subgroup $(\{0, 2, 4, 6, 8, 10\}, +_{12})$ must be normal. In D_3, the subgroup $\{e, \rho, \rho^2\}$ must be normal. In D_4, every four-element subgroup must be normal—you might like to pick one from the diagram in Section 7.3 and draw a table in which its elements appear together in the top left. As you do that, notice that every four-element subgroup contains the group's centre. Why is that?

For a further example, recall that S_n is the *symmetric group of order n*; its elements are permutations of the set $\{1, 2, 3, \ldots, n\}$ and its operation is composition. Section 6.9 noted that permutations can be constructed by combining *transpositions*—each of which swaps two elements—and that a permutation is *even* if it requires an even number of transpositions. The even permutations make up half of S_n and thus form a normal subgroup called the *alternating group A_n*. For instance, the elements of S_3 are the permutations

$$e \quad (12) \quad (13) \quad (23) \quad (123) \quad (132).$$

The permutation e requires zero transpositions, and $(123) = (13)(12)$ and $(132) = (12)(13)$ each require two. So $A_3 = \{e, (123), (132)\}$. Recalling that S_3 is isomorphic to D_3, you might want to imagine a triangle with vertices labelled $1, 2, 3$ and think about how A_3 corresponds to a subgroup of D_3. Does the subgroup A_4 of S_4 correspond to a subgroup of D_4? Why, or why not?

To consider relationships between D_4, A_4 and S_4, I find it helpful to adapt the previous section's vertical representation of the cosets of $3\mathbb{Z}$ in $\mathbb{Z}$ to construct similar diagrams for other groups and subgroups. The diagram below shows the right cosets of the subgroup $H = \{e, \rho, \rho^2, \rho^3\}$ in D_4, written in our original notation and in the generator-based notation from Section 6.8. The horizontal shift represents right composition with r_1 or r and the vertical shifts represent left composition with elements of H. What would change for the left cosets?

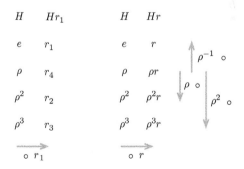

For $\{e, r_1\}$ as a non-normal subgroup of D_4, the left and right cosets and thus the columns differ.

$$
\begin{array}{cccc}
H & \rho H & \rho^2 H & \rho^3 H \\
e & \rho & \rho^2 & \rho^3 \\
r_1 & r_3 & r_2 & r_4
\end{array}
\quad \Big\downarrow \circ r_1
$$

$$\xrightarrow{\ \rho\, \circ\ }$$
$$\xrightarrow{\ \ \rho^2\, \circ\ \ }$$
$$\xrightarrow{\ \ \ \rho^3\, \circ\ \ \ }$$

$$
\begin{array}{cccc}
H & H\rho & H\rho^2 & H\rho^3 \\
e & \rho & \rho^2 & \rho^3 \\
r_1 & r_4 & r_2 & r_3
\end{array}
\quad \Big\downarrow r_1 \circ
$$

$$\xrightarrow{\ \circ\, \rho\ }$$
$$\xrightarrow{\ \ \circ\, \rho^2\ \ }$$
$$\xrightarrow{\ \ \ \circ\, \rho^3\ \ \ }$$

Now, recall that for $3\mathbb{Z}$ as a subgroup of $\mathbb{Z}$, the horizontal shifts represent remainders. So these diagrams help me to see an analogue of 'division with remainders' in dihedral groups and general groups.

- For $3\mathbb{Z}$ as a subgroup of $\mathbb{Z}$, left cosets have the form (remainder)$+3\mathbb{Z}$ and elements of cosets have the form (remainder)+(element of $3\mathbb{Z}$).
- For $H = \{e, \rho, \rho^2, \rho^3\}$ as a subgroup of D_4, left cosets have the form ('remainder')H and elements of cosets have the form ('remainder')(element of H).
- For H as a subgroup of G, left cosets have the additive or multiplicative forms ('remainder')$+H$ or ('remainder')H and elements of cosets have the form ('remainder')+(element of H) or ('remainder')(element of H).

$$
\begin{array}{ccccc}
H & g_1 H & g_2 H & \cdots & g_n H \\
e & g_1 & g_2 & \cdots & g_n \\
h_1 & g_1 h_1 & g_2 h_1 & \cdots & g_n h_1 \\
h_2 & g_1 h_2 & g_2 h_2 & \cdots & g_n h_2 \\
\vdots & \vdots & \vdots & & \vdots \\
h_m & g_1 h_m & g_2 h_m & \cdots & g_n h_m
\end{array}
$$

$$\xrightarrow{\ g_1\ }$$
$$\xrightarrow{\ \ g_2\ \ }$$

With this in mind, I made the diagram below to help myself understand the structure of S_4. The left column contains the elements of D_4, with rotations at the top and reflections at the bottom. The middle column contains these elements composed with the twist (12) (see Section 6.9). Which is performed first, the element of D_4 or the twist? What does the third column show? How do we know that this diagram shows every element of S_4? How many generators does this organization effectively use? Would you find an alternative organization more natural? Is D_4 a normal subgroup of S_4? Where are the elements of A_4? You might find that this diagram prompts other questions too.

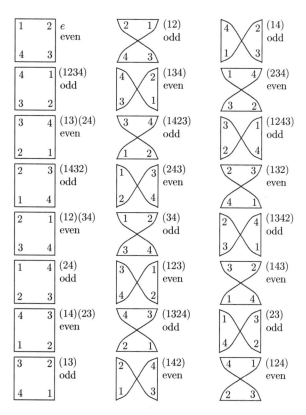

To conclude, a couple of classes of normal subgroups that we have not yet considered. First, the degenerate cases. A group G is always a subgroup of itself, and must be normal because for every $a \in G$, aG and Ga represent a whole row and column of the group table; they therefore include every element of G, so $aG = G = Ga$. Also, the identity alone is always a subgroup, and must be normal because for every $a \in G$, $a\{e\} = ae = a = ea = \{e\}a$. So every group has at least two normal subgroups. Second, students who have studied some Abstract Algebra might know that normal subgroups occur as *kernels* of *group homomorphisms*. If you already knew that, you might want to think about why. Either way, you can find more in Chapter 8.

7.7 Lagrange's Theorem

This final section broadens the discussion from normal subgroups to all subgroups, specifically to a property of subgroups captured in *Lagrange's Theorem*. Like all theorems, this can be stated in different ways. Section 3.5 included this simple version.

Lagrange's Theorem: Suppose that G is a finite group and H is a subgroup of G. Then the order of H divides the order of G.

We will start by considering meaning and logic. You might have considered meaning in Section 6.2, which asked whether $\mathbb{Z}_{12}$ could have subgroups of order five, or eight. Why does Lagrange's Theorem imply that it cannot? You might also have considered it in Section 6.7, which noted that a subgroup taking up more than half of a finite group must be the whole group. Why does Lagrange's Theorem imply that?

The logic of Lagrange's Theorem was discussed in Section 3.5, which observed that its converse is not true: the order of a subset might divide that of the group without that subset being a subgroup. A more subtle possible error is assuming that if a number divides the order of a group then there must exist a subgroup with that order. This is not true, but that is not obvious from the examples so far. For instance, the cyclic group $\mathbb{Z}_{12}$ has subgroups of orders $1, 2, 3, 4, 6$ and 12. The dihedral group D_4 has subgroups of orders $1, 2, 4$ and 8. The smallest group to provide a counterexample is A_4, which has order 12 but no subgroup of order 6. You might want to explore A_4 to investigate. You might also like to know that if a prime p divides the order of a group, then it *is* true that the group has a subgroup of order p. Do you have any feeling for why?

In fact, Lagrange's Theorem says more than the above; a fuller version involves cosets.

Lagrange's Theorem: Suppose that G is a finite group and H is a subgroup of G. Then the order of G is equal to the order of H multiplied by the number of cosets of H in G.

That is a bit wordy—you might prefer a more symbolic version using a new definition.

Definition: Let H be a subgroup of G. Then the **index** of H in G, denoted $|G:H|$, is the number of cosets of H in G.

Lagrange's Theorem: Suppose that G is a finite group and H is a subgroup of G. Then $|G| = |H||G:H|$.

Now, if you have read this chapter from the beginning, Lagrange's Theorem will not surprise you. Indeed, it might seem rather obvious. But, as everywhere in this book, much of the discussion has been about specific examples, which can provide insight but not general proof. So you might want to pause and think. Do we know enough to prove Lagrange's Theorem? Or is more work required?

A proof must establish that cosets 'fill up' the group without overlapping, and that each has the same order as H. Check that you agree. The filling-up part was established in Section 7.4, which proved that for any subgroup of any group, every element is in some coset and the cosets partition the group. The same-order part we have not yet considered. For the subgroups explored so far, every coset does have the same number of elements as the subgroup. But does that always hold?

One way to think about this intuitively is to imagine a row of a group table. Suppose that a subgroup H has elements denoted h_i, and consider the coset aH, which has elements ah_i. These elements are all in the same row and, because each row contains each group element exactly once, each ah_i must be different.

	g_1	h_1	g_2	h_2	g_3	g_4	...	h_i	h_j	g_n
...	...	...	...	...	...	...	...	...	...	...
a	ag_1	ah_1	ag_2	ah_2	ag_3	ag_4	...	ah_i	ah_j	ag_n
...	...	...	...	...	...	...	...	...	...	...

How can we capture that formally? With a subgroup H and coset aH, we want to prove that for $h_1, h_2 \in H$, if $h_1 \neq h_2$ then $ah_1 \neq ah_2$. Can you think of a way to do that? The idea is simple, but those negatives are difficult to handle, which makes it easier to remember that the contrapositive of a conditional statement is equivalent to that statement (see Section 3.5). Here, the contrapositive is: if $ah_1 = ah_2$ then $h_1 = h_2$. Notice where this is used in the lemma and proof below ('lemma' is a word often used for a 'small' or preparatory theorem).

Lemma: Suppose that G is a finite group, H is a subgroup of G, and $a \in G$. Then $|aH| = |H|$.

Proof: First, note that $aH = \{ah | h \in H\}$, so $|aH| \leq |H|$.

Now suppose that $\exists h_1, h_2 \in H$ such that $ah_1 = ah_2$.

Then, because a has an inverse $a^{-1} \in G$, $a^{-1}ah_1 = a^{-1}ah_2$.

So $h_1 = h_2$.

Thus if $h_1 \neq h_2$, then $ah_1 \neq ah_2$.

So $|aH| \not< |H|$ and we must have $|aH| = |H|$.

Now we can prove Lagrange's Theorem.

Lagrange's Theorem: Suppose that G is a finite group and H is a subgroup of G. Then $|G| = |H||G:H|$.

Proof: Suppose that G is a finite group with subgroup H.

Then the cosets of H in G partition G (Section 7.4).

And each coset has order $|H|$ (by the lemma above).

Thus $|G| = |H||G:H|$.

A short proof can make a theorem seem trivial, so remember how much reasoning is packed into the invoked theorem and lemma. Your course might do as we have, establishing those results first then presenting a short proof. Or it might present a long proof incorporating all of the underlying reasoning.

Now, Lagrange's Theorem has some immediate consequences. First, recall from Section 6.3 that the set generated by an element of a group is defined as below.

Definition: Let G be a group and $g \in G$. The **set generated by** g is
$$\langle g \rangle = \{g^n | n \in \mathbb{Z}\}.$$

The set $\langle g \rangle$ is always a cyclic subgroup of G (see Section 6.4). Its order is the smallest n such that g^n is the identity, and this n is defined to be the order of the *element g*. For instance, in D_4, the order of ρ is 4, the order of r_1 is 2, and the order of e is 1. Do keep your eye on object types: g is a group element, $\langle g \rangle = \{e, g, g^2, g^3, \ldots, g^n\}$ is a set, $|g|$ is a number, and $|\langle g \rangle|$ is the same number—why? With this in place, Lagrange's Theorem implies that the order of every element must divide the order of the group. You might want to consider more examples.

A second consequence of Lagrange's Theorem is that every group of prime order is cyclic. Why is that? Think about the definition of cyclic— see Section 6.3—and about what orders subgroups can take in a group of prime order.

A third consequence is this theorem.

Theorem: If a group G has order n, then $g^n = e$ for all $g \in G$.

Again, why must that be true? Every element's order divides n—how is that relevant? It is also worth considering the converse, which would say that if $g^n = e$ for all $g \in G$, then G has order n. This is not true: in the Klein four-group V (see Section 6.8), $g^2 = e$ for all $g \in V$, but V has order 4.

To conclude, some reasoning to show the power of abstract results. Consider S_4, for which $K = \{e, (12)(34), (13)(24), (14)(23)\}$ is a normal subgroup. We can prove that for every $g \in S_4$, $g^6 \in K$. Is that obvious to you? It is not to me—I am not familiar with this subgroup—and exploring it draws together some ideas from this chapter.

First, checking that K is normal is nontrivial but not difficult: we can find the left and right cosets of K in S_4 and check that they are the same. How would you do that efficiently? I might start by listing the subgroup, which will be one coset. Then, because the cosets partition the group, a second coset must be of the form aK where $a \notin K$. I might choose a simple element $a = (12) \in S_4$ and check that $(12)K = K(12)$.

K	$(12)K$		K	$K(12)$
e	(12)		e	(12)
$(12)(34)$	(34)		$(12)(34)$	(34)
$(13)(24)$	(1324)		$(13)(24)$	(1423)
$(14)(23)$	(1423)		$(14)(23)$	(1324)

A third coset must be of the form bK where $b \notin K \cup aK$ (the set $K \cup aK$ is the *union* of K and aK, the set of all elements that are in K or aK or both). So I might choose $b = (13)$. Can you finish this reasoning to show that the left and right cosets are equal? Do so and you will see that K is indeed normal in S_4.

K	$(12)K$	$(13)K$		K	$K(12)$	$K(13)$
e	(12)	(13)		e	(12)	(13)
$(12)(34)$	(34)	(1234)		$(12)(34)$	(34)	(1432)
$(13)(24)$	(1324)	(24)		$(13)(24)$	(1423)	(24)
$(14)(23)$	(1423)	(1432)		$(14)(23)$	(1324)	(1234)

How about the claim that for any g in S_4, $g^6 \in K$? This too is nontrivial but, in principle, not difficult. We could list all 24 elements of S_4, compose each with itself six times, and check that the result is in K. That, however, does not sound like fun. Fortunately, we can be much more efficient by applying theorems about quotient groups and group orders. Study the claim and proof below to see how.

Claim: Let $K = \{e, (12)(34), (13)(24), (14)(23)\} \trianglelefteq S_4$.

      Then for every $g \in S_4$, $g^6 \in K$.

Proof: Because K is normal, S_4/K is a quotient group.

      By Lagrange's Theorem, there are $24/4 = 6$ cosets of K in S_4.

      So the quotient group S_4/K has order 6.

      Let $g \in S_4$ be arbitrary and note that S_4/K has an element gK.

      Then, by the previous theorem, $(gK)^6$ is equal to the identity in S_4/K.

      But the identity in a quotient group is the subgroup, in this case K.

      So $(gK)^6 = K$.

      Now, $\forall a, b \in G$, $(aK)(bK) = (ab)K$,

      Thus $(gK)^6 = K \Rightarrow g^6 K = K$.

      So $g^6 \in K$ for every $g \in G$.

You might want to read this proof a few times to sort out the object types, thinking about group elements versus the cosets that form quotient group elements. But notice that the argument involves no calculations with specific group elements; rather, it relies on general results of Abstract Algebra. I do not expect you to have this kind of reasoning at your fingertips. But I would like you to understand the power of general reasoning and what you might aim for.

CHAPTER 8

Isomorphisms and Homomorphisms

This chapter explores and formalizes the notion that two groups can be isomorphic, meaning structurally identical. It introduces examples of isomorphisms and ways of establishing whether two groups are isomorphic or not. It then explores homomorphisms—maps between groups that respect their binary operations but are not necessarily bijective. Finally, it explains how the First Isomorphism Theorem links homomorphisms to normal subgroups and quotient groups.

8.1 What is an isomorphism?

Two groups are *isomorphic* if and only if they are structurally identical. For instance, $\mathbb{Z}_4$, the cyclic group of order four, is isomorphic to the multiplicative group $U_4 = (\{1, i, -1, -i\}, \times)$. Can you see this in their tables?

$+_4$	0	1	2	3
0	0	1	2	3
1	1	2	3	0
2	2	3	0	1
3	3	0	1	2

$\times$	1	i	-1	$-i$
1	1	i	-1	$-i$
i	i	-1	$-i$	1
-1	-1	$-i$	1	i
$-i$	$-i$	1	i	-1

Another group of order four is the Klein four-group V. This appears structurally different, as on the right below. But could e, a, b and c be re-ordered so that the table 'matches' that for $\mathbb{Z}_4$?

$+_4$	0	1	2	3
0	0	1	2	3
1	1	2	3	0
2	2	3	0	1
3	3	0	1	2

$*$	e	a	b	c
e	e	a	b	c
a	a	e	c	b
b	b	c	e	a
c	c	b	a	e

It could not, because structural properties of the two groups differ. Consider, for instance, orders of elements, defined as below (see also Section 7.7, and note that 'min' means 'minimum').

Definition: Let G be a group with identity e. Then the **order** of $g \in G$ is $\min\{n \in \mathbb{Z}^+ | g^n = e\}$. If no such n exists, g is said to have infinite order.

The group $\mathbb{Z}_4$ has operation $+_4$ and identity 0, so the element 1 has order four because $1 \neq 0$ and $1 +_4 1 \neq 0$ and $1 +_4 1 +_4 1 \neq 0$ but $1 +_4 1 +_4 1 +_4 1 = 0$. In V, every non-identity element has order 2 because $a * a = b * b = c * c = e$.

Now, the 'structural sameness' notion is intuitive rather than formal. It is completely correct, and there is nothing wrong with intuitive conceptions—experts use them routinely. Unfortunately, it is not that useful to work with, for two reasons. First, structural sameness is hard to see if elements do not line up conveniently in tables. Second, as with most intuitive notions, it cannot be used to construct formal mathematical arguments. Mathematicians do not argue that two groups are isomorphic by saying 'Look, they're the same!' (at least not in print—they might in conversation). And no one can use it to prove theorems.

So we need a formal way to work with isomorphisms, and this chapter will offer that and then go on to discuss *homomorphisms*. Isomorphisms and homomorphisms are closely related—unsurprisingly, given the names. Isomorphisms are 'tighter': they require exact sameness. Homomorphisms are 'looser': they require two groups to have specific common structure but not (necessarily) to be structurally identical. Some Abstract Algebra courses introduce homomorphisms first, which makes logical sense because these require fewer definitional properties. But I think that isomorphisms are so psychologically natural that it makes sense to start with those. We will get to homomorphisms in Section 8.6.

A first thing to note is the relationship between the words *isomorphic* and *isomorphism*. Two groups are *isomorphic* if and only if there is an *isomorphism* between them, where an isomorphism is a function 'matching' elements in one group to those in the other so as to respect their structures. For instance, $\mathbb{Z}_4$ is isomorphic to U_4 via the isomorphism ϕ (the Greek letter 'phi') represented below.

$+_4$	0	1	2	3
0	0	1	2	3
1	1	2	3	0
2	2	3	0	1
3	3	0	1	2

$$\xrightarrow{\phi}$$

$\phi(0) = 1$

$\phi(1) = i$

$\phi(2) = -1$

$\phi(3) = -i$

$\times$	1	i	-1	$-i$
1	1	i	-1	$-i$
i	i	-1	$-i$	1
-1	-1	$-i$	1	i
$-i$	$-i$	1	i	-1

It matters how ϕ is constructed because not all functions from $\mathbb{Z}_4$ to U_4 are isomorphisms. To see how restrictive the notion is, observe that a possible 'matching' function could map $0 \in \mathbb{Z}_4$ to any of $1, i, -1$ or $-i$. Then it could map $1 \in \mathbb{Z}_4$ to any of the remaining three elements, giving 4×3 options for 0 and 1 together. Extending this reasoning yields $4 \times 3 \times 2 \times 1 = 4! = 24$ potential matching functions from $\mathbb{Z}_4$ to U_4. How many respect the groups' structures and thus qualify as isomorphisms? Think this through, or take an educated guess.

One possible check involves element orders: if an element of one group has order n then so should its image under an isomorphism. This constrains isomorphisms pretty tightly. For instance, $\mathbb{Z}_4$ and U_4 each have one element of order one: their respective identities 0 and 1. So any isomorphism $\psi : \mathbb{Z}_4 \to U_4$ must have $\psi(0) = 1$ (ψ is the Greek letter 'psi'). Similarly, each group has one element of order two: 2 in $\mathbb{Z}_4$ because $2 +_4 2 = 0$, and -1 in U_4 because $(-1) \times (-1) = 1$. So any isomorphism must have $\psi(2) = -1$.

$+_4$	0	1	2	3
0	0	1	2	3
1	1	2	3	0
2	2	3	0	1
3	3	0	1	2

ψ
$\longrightarrow$

$\psi(0) = 1$

$\psi(2) = -1$

$\times$	1		-1	
1	1		-1	
		-1		1
-1	-1		1	
		1		-1

That leaves two elements of order four in each group: 1 and 3 in $\mathbb{Z}_4$ and i and $-i$ in U_4. The isomorphism ϕ had $\phi(1) = i$ and $\phi(3) = -i$. Could an isomorphism ψ have $\psi(1) = -i$ and $\psi(3) = i$ instead?

Inspecting full tables might help. Below, the elements are reordered on the right to show correspondences for the potential isomorphism ψ. Are these represented correctly in the header row and column and in the results of the operations $+_4$ and $\times$?

$+_4$	0	1	2	3
0	0	1	2	3
1	1	2	3	0
2	2	3	0	1
3	3	0	1	2

ψ
$\longrightarrow$

$\psi(0) = 1$
$\psi(1) = -i$
$\psi(2) = -1$
$\psi(3) = i$

$\times$	1	$-i$	-1	i
1	1	$-i$	-1	i
$-i$	$-i$	-1	i	1
-1	-1	i	1	$-i$
i	i	1	$-i$	-1

Everything works, so $\psi : \mathbb{Z}_4 \to U_4$ is an isomorphism. But it is the only alternative to ϕ, the only other map that respects the groups' structures. And this thinking is practical only because these groups are small. So easy-to-manipulate isomorphism criteria would be both mathematically desirable and practically useful. We need a definition.

8.2 Isomorphism definition

To motivate the definition of isomorphism, it is instructive to think beyond orders of elements to other properties that isomorphisms should respect. To do that, some notation helps. Suppose that G_1 and G_2 are groups and that $\phi : G_1 \to G_2$ is an isomorphism. Then every $g \in G_1$ maps to an element $\phi(g) \in G_2$.

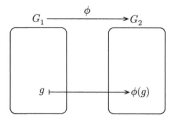

Now, in earlier mathematics, people often speak ambiguously about notation such as $\phi(g)$ or $f(x)$. They say things like 'the function $f(x)$', which is not badly misleading but not fully precise. In fact, f denotes the function, as in the notation $f: \mathbb{R} \to \mathbb{R}$ ('f from the reals to the reals'). The object x is an element of the domain, and $f(x)$ is its image, an element of the range or codomain. In our notation, ϕ is a function, g is an element of G_1, and $\phi(g)$ is an element of G_2. As ever, object types matter.

With an isomorphism $\phi: G_1 \to G_2$ in place, here are some questions.

- Suppose that G_1 is abelian (that its operation is commutative). What should be true about G_2?
- Suppose that e is the identity for G_1. What should be true about $\phi(e)$ in G_2? Can you relate that to orders of elements?
- If g has inverse g^{-1} in G_1, what should be the inverse of $\phi(g)$ in G_2?
- Suppose that $\langle g \rangle = G_1$ (that g generates G_1—see Section 6.3). What should be true about $\phi(g)$ in G_2? Can you relate this to orders of elements, too?

Happily, the answers to the questions are those that 'sound right'. Abelian groups should be isomorphic to abelian groups. Any isomorphism should map the identity in one group to the identity in the other, should map inverses to inverses, and should map generators to generators. That sounds like a fair few things to cover, but it turns out that two simple criteria force a map $\phi: G_1 \to G_2$ to respect all of these aspects of structure. The first criterion is:

$$\text{if } ab = c \in G_1, \text{ then } \phi(a)\phi(b) = \phi(c) \in G_2.$$

This forces what goes in the table 'on the right' to be correct with respect to structure. Do the tables below help you to see how it relates to commutativity, identities, inverses and generators? Some might be easier to think about than others.

	b				$\phi(b)$
a	c	$\xrightarrow{\phi}$	$\phi(a)$		$\phi(c)$

In formal work, it can be easier to shift attention from matching elements in isomorphic groups to isomorph*isms*, the functions that map between isomorphic groups. With that focus, the criterion can be stated as

$$\forall a,b \in G_1, \ \ \phi(ab) = \phi(a)\phi(b) \in G_2.$$

	b				$\phi(b)$
a	ab	$\xrightarrow{\phi}$	$\phi(a)$		$\phi(ab)$

The expression $\phi(ab) = \phi(a)\phi(b)$ requires disciplined thought, because it looks so right that people tend to assume that it always holds. One way to enforce that discipline is to recognize that in the table on the right, plenty of elements could occupy the $\phi(a)\phi(b)$ space; the criterion says that for ϕ to be an isomorphism, the only option is $\phi(ab)$. A second way is to remember that this general notation covers many possible groups with different operations. For instance, the operation in $\mathbb{Z}_4$ is $+_4$ and the operation in U_4 is $\times$. So, for the isomorphism $\phi : \mathbb{Z}_4 \to U_4$ from the previous section, the expression

$$\phi(ab) = \phi(a)\phi(b) \ \text{ actually reads } \ \phi(a +_4 b) = \phi(a) \times \phi(b).$$

That should make it seem less likely that the criterion would always hold. It also provides a link to a common algebraic error. In general, $(a+b)^2 \neq a^2 + b^2$ because the function $f \colon (\mathbb{R}, +) \to (\mathbb{R}, +)$ given by $f(x) = x^2$ is not an isomorphism: where $(a+b)^2 \neq a^2 + b^2$, we have $f(a+b) \neq f(a) + f(b)$.

A third way to enforce discipline is to think about orders of operations. In $\phi(ab)$, the elements $a, b \in G_1$ are first combined to give $ab \in G_1$. Then ϕ is applied to map $ab \in G_1$ to $\phi(ab) \in G_2$.

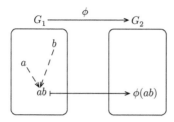

In $\phi(a)\phi(b)$, the function ϕ is first applied separately to map $a \in G_1$ to $\phi(a) \in G_2$ and $b \in G_1$ to $\phi(b) \in G_2$. Then these are combined to give $\phi(a)\phi(b) \in G_2$.

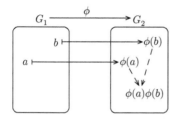

For functions in general, these two processes—combine then map, map then combine—need not give the same outcome. That's exactly what happens for $(a+b)^2 \neq a^2 + b^2$: adding then squaring need not be the same as squaring then adding. For isomorphisms, the criterion $\phi(ab) = \phi(a)\phi(b)$ means that they must have the same outcome.

Now, focusing on isomorphisms can feel unnatural because it introduces asymmetry. The relation 'is isomorphic to', denoted '$\cong$', is *symmetric*: if $G_1 \cong G_2$ then $G_2 \cong G_1$. So we can speak of isomorphic groups without privileging either one. But an isomorphism maps *from* one group *to* another; one is its domain and the other is its codomain. Fortunately, the asymmetry is only in what we write, because every isomorphism is invertible: if $G_1 \cong G_2$ via an isomorphism $\phi : G_1 \rightarrow G_2$, then $G_2 \cong G_1$ via an isomorphism $\phi^{-1} : G_2 \rightarrow G_1$.

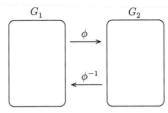

Invertibility brings us to the second isomorphism criterion. An isomorphism must be *bijective*; it must map one group to another 'perfectly', without mapping two elements to the same element or missing any out (people sometimes say that it must be 'one-to-one and onto'). Formally, a bijective function must be both *injective* and *surjective*, where these notions are defined and illustrated below.

Definition: A function $f: A \rightarrow B$ is **injective** if and only if $\forall a_1, a_2 \in A$, $f(a_1) = f(a_2) \Rightarrow a_1 = a_2$.

Definition: A function $f: A \rightarrow B$ is **surjective** if and only if $\forall b \in B$ $\exists a \in A$ such that $f(a) = b$.

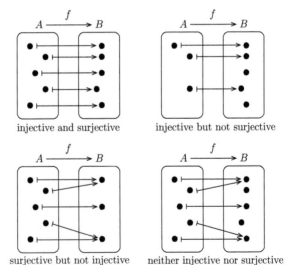

injective and surjective

injective but not surjective

surjective but not injective

neither injective nor surjective

Bijectivity is necessary for a function to be invertible—why? And it might be considered an obvious property of isomorphisms—two groups cannot be structurally identical if one has more elements than the other. If you therefore thought about bijectivity earlier, well done. If not, note that I implicitly assumed it when calculating the number of possible 'matching' functions from $\mathbb{Z}_4$ to U_4. How, exactly?

Now we have what we need to define *isomorphism* and *isomorphic groups*. Each definition can be formulated with a focus on the groups or a focus on the isomorphism. Here are versions that explicitly separate the two required properties.

Definition: Let G_1 and G_2 be groups. Then G_1 and G_2 are **isomorphic** if and only if there exists a function $\phi : G_1 \to G_2$ such that:

    1. ϕ is bijective;

    2. $\forall a, b \in G_1, \phi(ab) = \phi(a)\phi(b) \in G_2$.

Definition: Let G_1 and G_2 be groups. Then $\phi : G_1 \to G_2$ is an **isomorphism** if and only if:

    1. ϕ is bijective;

    2. $\forall a, b \in G_1, \phi(ab) = \phi(a)\phi(b) \in G_2$.

Separation, though, adds length. So you might see shorter versions.

Definition: Groups G_1 and G_2 are **isomorphic** if and only if there exists a bijection $\phi : G_1 \to G_2$ such that $\forall a, b \in G_1, \phi(ab) = \phi(a)\phi(b) \in G_2$.

Definition: $\phi : G_1 \to G_2$ is a **group isomorphism** if and only if ϕ is bijective and $\forall a, b \in G_1, \phi(ab) = \phi(a)\phi(b) \in G_2$.

Indeed, people often abbreviate further when notation is well established or when writing notes. Are you familiar with all of the abbreviations below? If not, can you work out what they mean?

Definition: $G_1 \cong G_2$ iff $\exists$ bijection $\phi : G_1 \to G_2$ s.t. $\forall a, b \in G_1, \phi(ab) = \phi(a)\phi(b)$.

Definition: $\phi : G_1 \rightarrow G_2$ is a **group isomorphism** iff ϕ is bijective and $\forall a, b \in G_1, \phi(ab) = \phi(a)\phi(b)$.

Which formulations do you prefer? Mathematicians value brevity, but they also tend to write carefully for students, so you might see longer versions. For yourself, you might write something shorter, though be sure not to lose the logic, and remember that there are two criteria. The expression '$\phi(ab) = \phi(a)\phi(b)$' tends to catch the eye because it uses algebraic symbols. The bijection property does not because it is expressed in a single word. Don't forget it.

8.3 Early isomorphism theory

So far, this chapter has played fast and loose with theory development, using intuitive ideas to suggest claims before considering definitions. This section will build theory in a logical order, starting with definitions and proving theorems. The theorem below captures the earlier observation about isomorphisms and abelian groups. What are its premises and conclusion (see Section 3.1)? Can you see a way to prove it?

Theorem: Suppose that G_1 and G_2 are isomorphic groups and that G_1 is abelian. Then G_2 is abelian.

The theorem has two premises and a conclusion, and to start a proof it is a good idea to write out what these mean in terms of definitions (see Section 3.7); to avoid confusion, it can be good to use different notation for general elements of G_1 and G_2.

Premise: G_1 and G_2 are isomorphic, meaning that there exists a bijection $\phi : G_1 \rightarrow G_2$ such that $\forall a, b \in G_1$, $\phi(ab) = \phi(a)\phi(b) \in G_2$.

Premise: G_1 is abelian, meaning that $\forall a, b \in G_1, ab = ba$.

Conclusion: G_2 is abelian, meaning that $\forall x, y \in G_2, xy = yx$.

We can then deduce things from the premises. Because $ab = ba \in G_1$, it must be true that $\phi(ab) = \phi(ba) \in G_2$. That would be true for any function, not just for an isomorphism. But, because ϕ is an isomorphism, we also have $\phi(ab) = \phi(a)\phi(b)$ and $\phi(ba) = \phi(b)\phi(a)$. Can you see how to glue this together to prove that G_2 is abelian? We can write a chain of equalities like this:

$$\phi(a)\phi(b) = \phi(ab) = \phi(ba) = \phi(b)\phi(a).$$

That might seem like enough because it demonstrates commutativity of the operation on $\phi(a), \phi(b) \in G_2$. And indeed this reasoning forms the crux of a valid proof. But it glosses over a subtlety: can everything in G_2 be expressed as $\phi(\text{something})$? The answer is yes, because ϕ is bijective. But, in a proof, it is polite to establish that general elements of G_2 can be written in this form. With that in mind, use the self-explanation training from Section 3.6 to study the theorem and proof below.

Theorem: Suppose that G_1 and G_2 are isomorphic groups and that G_1 is abelian. Then G_2 is abelian.

Proof: Suppose that $\phi : G_1 \to G_2$ is an isomorphism.

Suppose that G_1 is abelian.

Consider $x, y \in G_2$.

Because ϕ is bijective, $\exists a, b \in G_1$ such that $\phi(a) = x$ and $\phi(b) = y$.

Thus $xy = \phi(a)\phi(b)$

$\qquad = \phi(ab) \qquad$ because ϕ is an isomorphism

$\qquad = \phi(ba) \qquad$ because G_1 is abelian

$\qquad = \phi(b)\phi(a)$ because ϕ is an isomorphism

$\qquad = yx.$

Hence $\forall x, y \in G_2, xy = yx$.

So G_2 is abelian.

Did you read slowly enough? It is easy to be lax about self-explanation when a proof's steps seem simple. So maybe read again, concentrating on the justifications and on how the argument fits the theorem.

Next, we expect isomorphisms to map identities to identities. This can be proved using what I think of as 'one of those identity arguments', by which I mean a clever-trick thing using the fact that $ee = e$. Below is a theorem and proof. How does the notation capture which objects are elements of which groups?

Theorem: Suppose that G_1 and G_2 are groups and that $\phi : G_1 \to G_2$ is an isomorphism. Then $\phi(e_{G_1}) = e_{G_2}$.

Proof: By definition, $e_{G_1} = e_{G_1} e_{G_1}$.

So, because ϕ is an isomorphism,
$\phi(e_{G_1}) = \phi(e_{G_1} e_{G_1}) = \phi(e_{G_1})\phi(e_{G_1})$.
$\phi(e_{G_1}) \in G_2$ must have an inverse $(\phi(e_{G_1}))^{-1} \in G_2$.

So
$$\phi(e_{G_1}) = \phi(e_{G_1})\phi(e_{G_1})$$
$$\Rightarrow (\phi(e_{G_1}))^{-1}\phi(e_{G_1}) = (\phi(e_{G_1}))^{-1}\phi(e_{G_1})\phi(e_{G_1})$$
$$\Rightarrow e_{G_2} = e_{G_2}\phi(e_{G_1})$$
$$\Rightarrow e_{G_2} = \phi(e_{G_1}) \text{ as required.}$$

Would you write this proof differently? Would you, for instance, give the identities single-letter names? That would be less cumbersome, but would make it harder to track what was an element of what. Perhaps a good compromise would be to use e_1 and e_2 for the identities of G_1 and G_2. In the final line, would you reorder to write $\phi(e_{G_1}) = e_{G_2}$ instead of the other way around? Would you tag that onto the end, or reorder further up to make it come out that way? There are no right answers to these style questions, though your lecturer might have preferences. I would like you to be aware that mathematicians make choices in writing lecture notes and textbooks, and to think about advantages and disadvantages of different styles so that you can understand those choices. I would also like you to think about your own communication, and to be willing to rewrite something if you decide it would be better in alternative notation.

For inverses under isomorphisms, the natural theorem appears below.

Theorem: Suppose that G_1 and G_2 are groups and that $\phi : G_1 \to G_2$ is an isomorphism. Then $\forall a \in G_1, \phi(a^{-1}) = (\phi(a))^{-1} \in G_2$.

Again, notation is both friend and foe. The equality $\phi(a^{-1}) = (\phi(a))^{-1}$ 'looks right' so it is easy to remember, but it can be hard to see that there is anything to prove. Again it can help to think about orders of operations. In $\phi(a^{-1})$, we start with $a \in G_1$, take its inverse $a^{-1} \in G_1$, then apply ϕ to give $\phi(a^{-1}) \in G_2$.

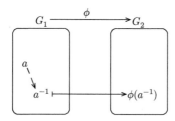

In $(\phi(a))^{-1}$, we start with $a \in G_1$, apply ϕ to give $\phi(a) \in G_2$, then take its inverse $(\phi(a))^{-1} \in G_2$.

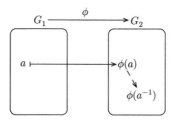

The theorem claims that if ϕ is an isomorphism, then these two processes—invert then map, map then invert—always give the same outcome. And the theorem is true, which we can think about via the previous result on identities: if a and b are mutually inverse in G_1, and the isomorphism maps the identity in G_1 to the identity in G_2, what do the following tables capture?

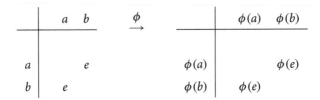

To construct a proof, we can again write down the premises and conclusion in terms of definitions. We might also consider other potentially useful properties involving inverses in G_1 and G_2.

Premise: $\phi : G_1 \to G_2$ is bijective.

Premise: $\forall a, b \in G_1, \phi(ab) = \phi(a)\phi(b)$.

Conclusion: $\forall a \in G_1, \phi(a^{-1}) = (\phi(a))^{-1}$.

Property: $\forall a \in G_1, \exists a^{-1} \in G_1$ such that $aa^{-1} = a^{-1}a = e_{G_1}$.

Property: $\forall x \in G_2, \exists x^{-1} \in G_2$ such that $xx^{-1} = x^{-1}x = e_{G_2}$.

Can you glue this information together? The theorem says that $\phi(a^{-1}) = (\phi(a))^{-1}$ for all $a \in G_1$, and a proof appears below. Which premises and properties does it use? Could you construct a proof differently?

Theorem: Suppose that G_1 and G_2 are groups and that $\phi : G_1 \to G_2$ is an isomorphism. Then $\forall a \in G_1, \phi(a^{-1}) = (\phi(a))^{-1} \in G_2$.

Proof: Let $a \in G_1$.

Then $\exists a^{-1} \in G_1$ such that $aa^{-1} = a^{-1}a = e_{G_1}$.

So $\phi(aa^{-1}) = \phi(a^{-1}a) = \phi(e_{G_1})$.

Now ϕ is a homomorphism so:

$\phi(aa^{-1}) = \phi(a)\phi(a^{-1})$ and $\phi(a^{-1}a) = \phi(a^{-1})\phi(a)$
and $\phi(e_{G_1}) = e_{G_2}$.

Thus $\phi(a)\phi(a^{-1}) = \phi(a^{-1})\phi(a) = e_{G_2}$, i.e. $(\phi(a))^{-1} = \phi(a^{-1})$.

Hence $\forall a \in G_1, \phi(a^{-1}) = (\phi(a))^{-1}$.

Finally, I noted that we expect isomorphisms to map generators of one group to generators of the other. This is a bit fiddlier, and your course will probably provide a theorem and proof. But the reasoning strategies suggested here should help, and I encourage you to try proving this yourself.

8.4 Example isomorphisms

The previous section developed some theory, and we could keep building on that. But, for me, there is tension between building theory and feeling that I could 'climb back down' to the examples to which that theory applies. For some people this is not an issue: they are confident in abstract reasoning and satisfied to see theory grow, especially if it is tidy. I like tidiness too—that is why I prefer pure to applied mathematics—but I start to feel vertiginous if I go too far without 'grounding' myself.

Another reason to study examples is that you might be wondering whether isomorphisms are, in fact, a bit boring. Once you've noticed that two groups are structurally identical, it might seem that there isn't much more to say. Yes, their identities and inverses match, but that is pretty obvious. And working with small groups might make isomorphisms seem trivial. For instance, cyclic groups can be understood as points evenly spaced around a circle, so it is not surprising that they match other groups with that feature (see Section 6.6).

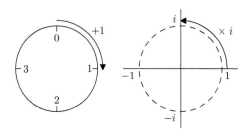

But we have also seen functions that are not isomorphisms. The function $f: (\mathbb{R}, +) \to (\mathbb{R}, +)$ given by $f(x) = x^2$ is not an isomorphism and, of the 24 bijections from $\mathbb{Z}_4$ to U_4, only two are isomorphisms. So there is a sense in which 'most' functions are not isomorphisms.

And many interesting isomorphisms are less obvious. In Section 6.8, you might have observed that the symmetry groups of a non-square rectangle and a rhombus are isomorphic to the Klein four-group V; so is the quotient group $D_4/\{e, \rho^2\}$ from Section 7.3. Other isomorphisms exist between groups of numbers and groups of transformations. For instance, translations along a line form a group under composition—why? Check against the group axioms. Moreover, if we denote a translation a units to the right by t_a and let $T = \{t_a | a \in \mathbb{R}\}$ (the set of all such translations), then $(T, \circ)$ and $(\mathbb{R}, +)$ are isomorphic via the map

$$\phi : (\mathbb{R}, +) \to (T, \circ) \text{ given by } \phi(a) = t_a.$$

This ϕ is an isomorphism because it is bijective—every real number corresponds to a unique translation and vice versa—and because it respects the groups' operations. For instance, $6 + (-2) = 4$ in $(\mathbb{R}, +)$ corresponds to $t_6 \circ t_{-2} = t_4$ in $(T, \circ)$.

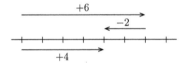

For the general case, check that you believe every equality in this chain:

$$\phi(a + b) = t_{a+b} = t_a \circ t_b = \phi(a) \circ \phi(b).$$

That example might seem obvious if you were brought up on number lines. But how must numbers have 'felt' to people in the past who used only natural numbers and recorded them using tally marks? The psychology must have changed a lot, I think.

Isomorphisms also exist for other sets of transformations. For instance, consider the set of all enlargements or *dilations* of the plane centred at $(0,0)$. These form a group under composition—again, why? If d_a denotes dilation by a factor of a and $D = \{d_a | a \in \mathbb{R}\}$, does $(D, \circ)$ also form a group isomorphic to $(\mathbb{R}, +)$? No, it does not even form a group, because d_0 has no inverse. We can restrict so that, instead, $D = \{d_a | a \in \mathbb{R} \setminus \{0\}\}$, but even then, $(D, \circ)$ is not isomorphic to $(\mathbb{R} \setminus \{0\}, +)$

because the dilations form a multiplicative rather than an additive structure. For instance, dilating by a factor of 6 then by a factor of $\frac{1}{2}$ yields a dilation by a factor of 3. I think of this as in the diagram below, but the more algebraically minded might simply note that for every point (x, y) in the plane, $\frac{1}{2} \cdot 6(x, y) = \frac{1}{2}(6x, 6y) = (3x, 3y)$.

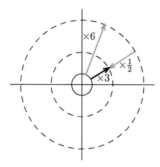

Did you notice that $\mathbb{R} \backslash \{0\}$ includes negative numbers? What is the effect of d_a if a is negative? For any point (x, y), $-1(x, y) = (-x, -y)$, so d_{-1} is a 180° rotation—check that you believe this. Similarly, $-3(x, y) = (-3x, -3y)$, so d_{-3} is a 180° rotation with dilation by a factor of 3. What is $d_{-3} \circ d_{-1}$? In general, $(D, \circ)$ is isomorphic to $(\mathbb{R} \backslash \{0\}, \times)$ via

$$\phi : (\mathbb{R} \backslash \{0\}, \times) \to (D, \circ) \text{ given by } \phi(a) = d_a,$$

which is an isomorphism because it is bijective and because

$$\phi(a \times b) = d_{a \times b} = d_a \circ d_b = \phi(a) \circ \phi(b).$$

Finally, consider rotations of the plane centred at $(0, 0)$. These form a group under composition—again, why? If ρ_a denotes an anticlockwise rotation through a radians and $X = \{\rho_a | a \in \mathbb{R}\}$, does $(X, \circ)$ form a group isomorphic to $(\mathbb{R}, +)$, or to $(\mathbb{R} \backslash \{0\}, \times)$? Its structure is additive because, for instance, rotating through $a + b$ is equivalent to rotating through a then through b, so $\rho_{a+b} = \rho_a \circ \rho_b$. Thus the map $\phi : (\mathbb{R}, +) \to (X, \circ)$ respects the two groups' operations because

$$\phi(a + b) = \rho_{a+b} = \rho_a \circ \rho_b = \phi(a) \circ \phi(b).$$

But is ϕ a bijection? No: the link between these structures is more complex because ρ_a and $\rho_{a+2\pi}$ have the same effect on every point in the plane, so $\phi(a) = \phi(a+2\pi)$. Thus ϕ is not an isomorphism (though it is a homomorphism—more on this in Section 8.7).

Turning from transformations to numbers, if $\mathbb{R}^+ = \{x \in \mathbb{R}|x > 0\}$ then the function $\phi : (\mathbb{R}, +) \rightarrow (\mathbb{R}^+, \times)$ given by $\phi(x) = e^x$ is an isomorphism. You might never have thought about it in this way, but the exponential function has the required properties. First, check the domain and image: for every $x \in \mathbb{R}$, $e^x \in \mathbb{R}^+$. Then check bijectivity: ϕ is injective and surjective because no two values of $x \in \mathbb{R}$ map to the same element of $\mathbb{R}^+$ and every element of $\mathbb{R}^+$ is 'hit'. Finally, ϕ is an isomorphism because

$$\phi(x + y) = e^{x+y} = e^x e^y = \phi(x)\phi(y).$$

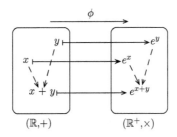

Familiarity with the exponential function might help to ground the preceding theory. What are the identities in $(\mathbb{R}, +)$ and $(\mathbb{R}^+, \times)$ and how does ϕ link those? How do inverses work for each group, and how does ϕ link those? Familiarity might also clarify that isomorphisms are general versions of precisely this kind of matching: in $(\mathbb{R}, +)$ and $(\mathbb{R}^+, \times)$, the sets and operations differ but the structures are identical.

That said, the idea that $(\mathbb{R}, +)$ and $(\mathbb{R}^+, \times)$ are isomorphic might be confusing because these structures do not seem literally 'the same'. I think this gets to the heart of what isomorphisms are. The groups $(\mathbb{R}, +)$ and $(\mathbb{R}^+, \times)$ are structurally identical in that there is a bijection between them that respects their operations. But we are not accustomed to thinking of $\mathbb{R}$ as a group with only an additive operation, or $\mathbb{R}^+$ as a group with only a multiplicative operation. We know much more about the real numbers, and it is hard to 'forget' that knowledge. Views of the reals that consider more operations are discussed in Chapter 9.

For now, you might like to know that ϕ and its inverse given by $\phi^{-1}(x) = \ln(x)$ used to have serious practical utility. When calculation was done by hand, multiplying was hard. Would you want to calculate $4,566,789 \times 132,453$ using long multiplication? Adding is easier, and calculations can be done by looking up logarithms, adding those, then taking exponents.

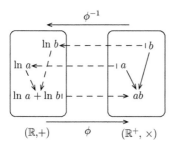

To conclude this section, an example using symmetries. I learned while writing this book that the group of rotations of a cube is isomorphic to S_4, the the set of permutations of $\{1, 2, 3, 4\}$ under composition (see Section 6.9). Does that seem plausible? Is it obvious? For me, plausibility is fine. The group S_4 has $4! = 24$ elements, and cubes have 6 faces, 8 vertices and 12 edges—all of those numbers 'go with' 24. But it is not obvious, so I can try to count. There is the identity rotation, obviously, plus three non-identity rotations about each of three distinct face-centred axes. That is ten so far. Then there is one rotation through each of the six axes joining the midpoints of opposite edges. That takes us to 16. Finally, there are two non-identity rotations about each of the four diagonals that join opposite vertices. That gives 24 in total.

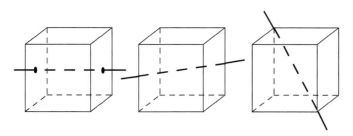

Of course, counting tells us nothing about the results of composing these rotations. For that, you can experiment, or we can reach for theory and use the fact that, as I learned, each rotation of the cube can be thought of as a unique permutation of the diagonals that join opposite vertices in pairs. There are four such diagonals, so this approach links more directly to S_4. Which permutations of $\{1, 2, 3, 4\}$ correspond to which rotations?

8.5 Isomorphic or not?

Because identical structures occur in different areas of mathematics, an obvious question is which groups are isomorphic to which others. This means that students are often asked to work out whether pairs of groups are isomorphic or not. What would be your approach to that? Try to put it into words before reading on.

Sometimes it is easy to tell that two groups are *not* isomorphic, because isomorphisms are bijections so isomorphic groups must have the same orders. This means that $\mathbb{Z}_4$ cannot be isomorphic to D_3 (the group of symmetries of an equilateral triangle—see Section 6.7) because $\mathbb{Z}_4$ has four elements and D_3 has six. But the group $\mathbb{Z}_6$ has six elements—is it isomorphic to D_3? Can you justify your answer?

When attempting such questions, students sometimes create work for themselves by being insufficiently strategic. They look at the definition of *isomorphic*, as below.

Definition: Groups G_1 and G_2 are **isomorphic** if and only if there exists a bijection $\phi : G_1 \to G_2$ such that $\forall a, b \in G_1, \phi(ab) = \phi(a)\phi(b) \in G_2$.

This lists the bijection criterion first, so they try to find isomorphisms by constructing bijections. That is not a good strategy because, as we have seen, the number of possible bijections can be huge. For instance, there are $6! = 720$ distinct bijections from $\mathbb{Z}_6$ to D_3. It would take extreme good luck to stumble upon an isomorphism, or extreme patience to check that none is an isomorphism.

A better strategy is to use theorems about properties that isomorphic groups share. For instance, $\mathbb{Z}_6$ and D_3 are not isomorphic because $\mathbb{Z}_6$ is cyclic and D_3 is not, or because $\mathbb{Z}_6$ is abelian and D_3 is not, or because $\mathbb{Z}_6$ has two elements of order 6 and D_3 has none. Property observations are particularly useful for infinite groups. For instance, are $GL(2, \mathbb{R})$ and $(\mathbb{R}\backslash\{0\}, \times)$ isomorphic? My intuition says no because the matrix group feels infinite 'in a different way'. But that does not make a good mathematical argument. However, we can deduce that they are not isomorphic because $(\mathbb{R}\backslash\{0\}, \times)$ is abelian and $GL(2, \mathbb{R})$ is not.

Of course, if we happen to know that two infinite sets have different *cardinalities*—different 'numbers of elements'—we can use that. For instance, $(\mathbb{R}, +)$ is not isomorphic to $(\mathbb{Q}, +)$ because there is no bijection between $\mathbb{R}$ and $\mathbb{Q}$. There *is* a bijection between $\mathbb{N}$ and $\mathbb{Q}$, which surprises most people because there seem to be 'a lot more' rational than natural numbers. A visual way to demonstrate that a bijection exists is to arrange the elements of $\mathbb{Q}$ as below, where all rationals would be included in the infinite array. Then follow the arrow, setting $\phi(1)$ to be the first rational it 'hits', $\phi(2)$ to be the second, and so on, ignoring the repeats (for instance, -1 is 'counted' so $-\frac{2}{2}$ is ignored).

$\phi(1) = 0 \quad \phi(2) = 1 \quad \phi(3) = \frac{1}{2} \quad \phi(4) = -\frac{1}{2} \quad \phi(5) = -1 \quad \phi(6) = -2\ldots$

This defines a bijection because ϕ matches each natural to a rational without leaving any out or sending two naturals to the same rational.

Is ϕ an isomorphism? No. The operation in both groups is addition, so an isomorphism would require that for every $a, b \in \mathbb{N}$, $\phi(a + b) = \phi(a) + \phi(b) \in \mathbb{Q}$. But, for instance,

$$\phi(2 + 3) = \phi(5) = -1 \neq 1 + \tfrac{1}{2} = \phi(2) + \phi(3).$$

Could there be a different bijection between $(\mathbb{N}, +)$ and $(\mathbb{Q}, +)$ that is an isomorphism? Again, my intuition says no. But it would have said no to the existence of a bijection, so perhaps it is not reliable for infinite groups. Fortunately, we can relate possible isomorphisms to equations. In $(\mathbb{N}, +)$, the equation $x + x = 1$ has no solution. But an isomorphism $\phi : (\mathbb{N}, +) \to (\mathbb{Q}, +)$ would require that $\phi(x) + \phi(x) = \phi(1)$ and, whatever the value of $\phi(1) \in \mathbb{Q}$, this equation would have a solution $\phi(x) \in \mathbb{Q}$. So the structures cannot be the identical and $(\mathbb{N}, +) \not\cong (\mathbb{Q}, +)$. Similarly, $(\mathbb{R} \backslash \{0\}, \times)$ is not isomorphic to $(\mathbb{C} \backslash \{0\}, \times)$ because in $(\mathbb{R} \backslash \{0\}, \times)$, the equation $x \times x = -1$ has no solution. How would the argument go?

To round off this discussion, it is worth revisiting the two isomorphisms from $\mathbb{Z}_4$ to U_4. These can be reconceived as isomorphisms from $\mathbb{Z}_4$ to itself, as *automorphisms*. One, $\phi : \mathbb{Z}_4 \to \mathbb{Z}_4$, is the identity isomorphism that maps each element to itself. The other, $\psi : \mathbb{Z}_4 \to \mathbb{Z}_4$, is represented below. You might want to check that this corresponds to ψ from Section 8.1.

$+_4$	0	1	2	3
0	0	1	2	3
1	1	2	3	0
2	2	3	0	1
3	3	0	1	2

ψ
$\xrightarrow{}$
$\psi(0) = 0$
$\psi(1) = 3$
$\psi(2) = 2$
$\psi(3) = 1$

$+_4$	0	3	2	1
0	0	3	2	1
3	3	2	1	0
2	2	1	0	3
1	1	0	3	2

These are the only automorphisms on $\mathbb{Z}_4$, because Section 8.1's argument about orders of elements applies. But we can also argue using generators. The group $\mathbb{Z}_4$ is cyclic so a single element generates the whole group: both 1 and 3 are generators. Under an isomorphism, each must map to a generator. So 1 must map to 1 or to 3. And once we know where a generator goes, the isomorphism is completely determined. Here, for instance,

$$\psi(1) = 3$$
$$\Rightarrow \psi(2) = \psi(1+1) = \psi(1) + \psi(1) = 3 + 3 = 2$$
$$\Rightarrow \psi(3) = \psi(2+1) = \psi(2) + \psi(1) = 2 + 3 = 1$$
$$\Rightarrow \psi(0) = \psi(3+1) = \psi(3) + \psi(1) = 1 + 3 = 0.$$

Can you link this to generators for other cyclic groups, as discussed in Section 6.4? How many automorphisms are there on $\mathbb{Z}_5$? On $\mathbb{Z}_{12}$? On $\mathbb{Z}$?

For practice in climbing through levels of abstraction, we can imagine a group and think about its automorphisms; on $\mathbb{Z}_4$, there are two automorphisms ϕ and ψ. Each automorphism is a function—indeed, a bijection—sending the set to itself. So automorphisms can be combined using composition. Composing two automorphisms gives another, function composition is always associative, there is an identity automorphism (sending every element to itself), and each automorphism has an inverse (sending every element back to where it came from). The automorphisms under composition therefore form a 'higher-level' group; for $\mathbb{Z}_4$, the group of automorphisms is isomorphic to $\mathbb{Z}_2$.

$\circ$	ϕ	ψ
ϕ	ϕ	ψ
ψ	ψ	ϕ

When people use the word *abstract* in Abstract Algebra, they are not messing around.

8.6 Homomorphisms

We turn now from isomorphisms to homomorphisms. The beginning of this chapter noted that these are closely related, but homomorphisms are 'looser' in that they have fewer definitional properties. In fact, we simply drop the bijection criterion but keep the 'respect the operations' criterion.

Definition: $\phi : G_1 \to G_2$ is a **group homomorphism** if and only if $\forall a, b \in G_1, \phi(ab) = \phi(a)\phi(b) \in G_2$.

This means that every isomorphism is a homomorphism but not every homomorphism is an isomorphism. Nevertheless, some existing theory applies. Review Section 8.3 and you will see that the proofs about identities and inverses under isomorphisms do not use the bijection criterion. So they apply to yield equivalent theorems about homomorphisms.

Theorem: Suppose that G_1 and G_2 are groups and that $\phi : G_1 \to G_2$ is a homomorphism. Then $\phi(e_{G_1}) = e_{G_2}$.

Theorem: Suppose that G_1 and G_2 are groups and that $\phi : G_1 \to G_2$ is a homomorphism. Then $\forall a \in G_1, \phi(a^{-1}) = (\phi(a))^{-1}$.

Does a homomorphism send abelian groups to abelian groups and generators to generators? Not necessarily. Section 8.3 established that if $G_1 \cong G_2$ and G_1 is abelian, so is G_2. But the proof used the fact that isomorphisms are surjective. A homomorphism might not be surjective—it might 'miss out' some elements of G_2. So elements in the *image* of ϕ (often denoted 'im ϕ') will commute with one another, but the proof says nothing about elements outside im ϕ.

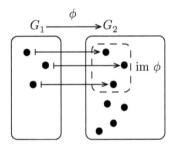

Non-surjective homomorphisms can also help in thinking about generators. A natural non-surjective homomorphism occurs when G_1 is a subgroup of G_2. For instance, $(3\mathbb{Z}, +)$ is a subgroup of $(\mathbb{Z}, +)$, and the map $\phi : (3\mathbb{Z}, +) \to (\mathbb{Z}, +)$ given by $\phi(x) = x$ maps $3\mathbb{Z}$ to a copy of itself inside $\mathbb{Z}$.

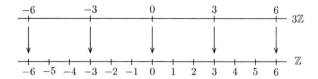

The map ϕ is not surjective: many elements of $\mathbb{Z}$ are not in $\operatorname{im}\phi$. But it is a homomorphism because for every $a, b \in 3\mathbb{Z}$, $\phi(a+b) = a + b = \phi(a) + \phi(b)$ (the operation in both groups is addition). And ϕ is injective, so it is called a *monomorphism*. A similar inclusion monomorphism occurs for every subgroup of every group: if H is a subgroup of G, there is a natural monomorphism $\phi : H \to G$ given by $\phi(h) = h$. Do these monomorphisms map generators to generators? Not necessarily. For instance, one generator of $3\mathbb{Z}$ is 3, but its image for the map above does not generate $\mathbb{Z}$.

Could an alternative homomorphism $\psi : (3\mathbb{Z}, +) \to (\mathbb{Z}, +)$ map a generator to a generator? The obvious map to try is that defined by $\psi(x) = x/3$, so that $\psi(3) = 1$. Is ψ a homomorphism? Yes, because for every $a, b \in 3\mathbb{Z}$, $\psi(a+b) = (a+b)/3 = a/3 + b/3 = \psi(a) + \psi(b)$. Indeed, ψ is bijective so it is actually an isomorphism, one that 'squashes everything in'.

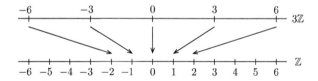

Homomorphisms can also 'spread everything out'. For instance, a homomorphism $\phi : (\mathbb{Z}, +) \to (\mathbb{Q}, +)$ could map $1 \in \mathbb{Z}$ to $1 \in \mathbb{Q}$, so $\phi(1) = 1$. But suppose, instead, that $\phi(1) = 2$. Then

$$\phi(1) = 2$$
$$\Rightarrow \phi(2) = \phi(1+1) = \phi(1) + \phi(1) = 2 + 2 = 4$$
$$\Rightarrow \phi(3) = \phi(2+1) = \phi(2) + \phi(1) = 4 + 2 = 6$$
$$\Rightarrow \phi(4) = \phi(3+1) = \phi(3) + \phi(1) = 6 + 2 = 8, \text{ etc.}$$

Also, the theorem about identities means that $\phi(0) = 0$, and the theorem about inverses translates into additive form as $\forall a \in G_1$, $\phi(-a) = -(\phi(a))$. Thus, knowing where the generator goes specifies the whole map.

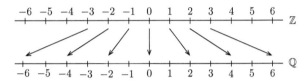

What happens if $\phi(1) = \frac{1}{10}$? Does every homomorphism $\phi : (\mathbb{Z}, +) \to (\mathbb{Q}, +)$ squash everything in or spread everything out? What if $\phi(1) = 0$?

Next, how do these ideas play out in finite groups? For instance, what homomorphisms exist from $\mathbb{Z}_6$ to $\mathbb{Z}_{12}$, or to $\mathbb{Z}_5$, or to itself? From $\mathbb{Z}_6$ to $\mathbb{Z}_{12}$, there is a monomorphism with $\phi(1) = 2$. What other homomorphisms exist? From $\mathbb{Z}_6$ to $\mathbb{Z}_5$ there are no nontrivial homomorphisms. Does that surprise you? Perhaps not, because 6 and 5 are coprime, so their structures do not 'go well together'. To understand formally, observe that the generator $1 \in \mathbb{Z}_6$ must map to something in $\mathbb{Z}_5$, and that most possibilities lead to contradictions. For instance, setting $\phi(1) = 1 \in \mathbb{Z}_5$ gives the following—what goes wrong?

$$\phi(1) = 1$$
$$\Rightarrow \phi(2) = \phi(1 +_6 1) = \phi(1) +_5 \phi(1) = 1 +_5 1 = 2$$
$$\Rightarrow \phi(3) = \phi(2 +_6 1) = \phi(2) +_5 \phi(1) = 2 +_5 1 = 3$$
$$\Rightarrow \phi(4) = \phi(3 +_6 1) = \phi(3) +_5 \phi(1) = 3 +_5 1 = 4$$
$$\Rightarrow \phi(5) = \phi(4 +_6 1) = \phi(4) +_5 \phi(1) = 4 +_5 1 = 0$$
$$\Rightarrow \phi(0) = \phi(5 +_6 1) = \phi(5) +_5 \phi(1) = 0 +_5 1 = 1$$
$$\Rightarrow \phi(1) = \phi(0 +_6 1) = \phi(0) +_5 \phi(1) = 1 +_5 1 = 2$$

The calculations yield two distinct values for $\phi(1)$, so ϕ is not a function, never mind a homomorphism. Similar contradictions arise for most elements of $\mathbb{Z}_5$ (check), but one option works: setting $\phi(1) = 0$ yields

$$\phi(1 +_6 1) = \phi(1) +_5 \phi(1) = 0 +_5 0 = 0$$

and, by extension, $\phi(a) = 0$ for every $a \in \mathbb{Z}_6$. This defines a homomorphism because for every $a, b \in \mathbb{Z}_6$, $\phi(a +_6 b) = 0 = 0 +_5 0 = \phi(a) +_5 \phi(b)$. And this reasoning generalizes: for any two groups, there is a degenerate or *trivial* homomorphism mapping every element of the first to the identity of the second. This means that there is at least one homomorphism between any two groups. So, unlike saying that two groups are isomorphic, saying that two groups are 'homomorphic' would be completely uninformative.

How about homomorphisms from $\mathbb{Z}_6$ to itself? Run through some calculations and you will see that homomorphisms 'organize' elements of $\mathbb{Z}_6$ into sets. For instance, a homomorphism $\phi : \mathbb{Z}_6 \to \mathbb{Z}_6$ with $\phi(1) = 3$ must have

$$\phi(2) = \phi(1 +_6 1) = \phi(1) +_6 \phi(1) = 3 +_6 3 = 0,$$

and so on. So it organizes the elements as in the diagram below, 'collapsing' the group to the subgroup $(\{0, 3\}, +_6)$. If you have read Chapter 7 then you might notice the pattern in what gets collapsed to where: the sets $\{0, 2, 4\}$ and $\{1, 3, 5\}$ are the cosets of the subgroup $(\{0, 2, 4\}, +_6)$ in $\mathbb{Z}_6$. This is not a coincidence, and the general phenomenon is explored in Section 8.7.

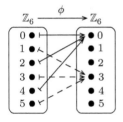

This notion of organizing or collapsing is informal but general—it can be used to understand other homomorphisms. Consider, for instance, an *evaluation homomorphism* ϕ mapping each continuous function from $\mathbb{R}$ to $\mathbb{R}$ to a single real number by evaluating it at a specific value, say $x = 2$.

This ϕ can be interpreted as collapsing each function to its value at $x = 2$, or organizing functions into sets by their values at $x = 2$. For instance, ϕ maps both f_1 and f_2 below to the number 4.

$f_1 : \mathbb{R} \to \mathbb{R}$ defined by $f_1(x) = 3x - 2$, then $\phi(f_1) = f_1(2) = 4$;

$f_2 : \mathbb{R} \to \mathbb{R}$ defined by $f_2(x) = x^2$, then $\phi(f_2) = f_2(2) = 4$.

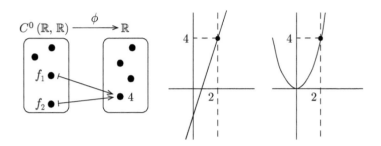

It might seem weird to write things like $\phi(f)$. But thinking at this level allows us to understand $\phi : (C^0(\mathbb{R}, \mathbb{R}), +) \to (\mathbb{R}, +)$ as a homomorphism, because for every $f, g \in C^0(\mathbb{R}, \mathbb{R})$,

$$\phi(f + g) = (f + g)(2) = f(2) + g(2) = \phi(f) + \phi(g).$$

Another homomorphism maps matrices to determinants. You might know that if A and B are matrices with determinants $\det(A)$ and $\det(B)$, then $\det(AB) = \det(A)\det(B)$. This is precisely the homomorphism criterion, although it is worth considering the groups and map. In our notation, G_1 is a group of matrices, say $GL(2, \mathbb{R})$, under matrix multiplication.

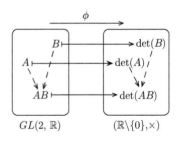

The map is 'det', which collapses each matrix to its determinant, or organizes matrices into sets according to their determinants. The group G_2 is $(\mathbb{R}\backslash\{0\}, \times)$, because every invertible matrix has a nonzero determinant and the equality $\det(AB) = \det(A)\det(B)$ uses multiplication in $\mathbb{R}$.

Finally, there is a homomorphism $\phi : (\mathbb{R}, +) \rightarrow (U, \times)$, where U is the unit circle $U = \{x \in \mathbb{C} \,|\, |x| = 1\}$ (see Section 6.6). This might seem unlikely because U is unlike $\mathbb{R}$—it 'goes round and round', not 'on forever in a straight line'. But recall that we can imagine wrapping a bendy integer number line around a clock to line up numbers that are congruent modulo 12—see Section 3.3. Similarly, we can imagine wrapping a bendy real number line around a circle. How should the 'lining up' work? The theory of homomorphisms provides a steer: homomorphisms send identities to identities, so any homomorphism $\phi : (\mathbb{R}, +) \rightarrow (U, \times)$ must have $\phi(0) = 1$.

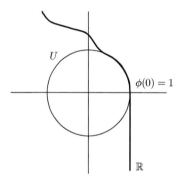

But how 'tight' should the winding be? What else should map to 1? There are two natural possibilities. One is to keep the 'scales' the same, in which case the set of numbers mapping to 1 would be $\{\ldots, -4\pi, -2\pi, 0, 2\pi, 4\pi, \ldots\}$. However, the homomorphic structures are more readily apparent by defining the map $\phi : (\mathbb{R}, +) \rightarrow (U, \times)$ by $\phi(x) = e^{2\pi i x}$. Putting 2π 'in' the map winds the number line around the circle once every integer, so that every integer maps to 1. And ϕ is a homomorphism because for every $x, y \in \mathbb{R}$,

$$\phi(x + y) = e^{2\pi i(x+y)} = e^{2\pi i x}e^{2\pi i y} = \phi(x)\phi(y).$$

This map effectively collapses real numbers 'modulo 1', organizing them into positions on the circle according to their 'remainders' on division by 1. This idea will be formalized in the next section.

8.7 The First Isomorphism Theorem

This final section links homomorphisms to quotient groups via the *First Isomorphism Theorem*.[1] If you like mathematical theory that draws ideas together, you will like this—it formalizes the idea that homomorphisms organize domain elements into sets that 'collapse' to image elements. Those sets turn out to be cosets of normal subgroups and thus to form elements of quotient groups (please read Chapter 7 if you do not already know about these concepts). A key focus is the set of elements that map to the identity, which is called the *kernel* of the homomorphism.

Definition: Suppose that G_1 and G_2 are groups and that $\phi : G_1 \rightarrow G_2$ is a homomorphism. Then the **kernel** of ϕ is $\ker \phi = \{k \in G_1 | \phi(k) = e_{G_2}\}$.

In words, the kernel of a homomorphism $\phi : G_1 \rightarrow G_2$ is the subset of G_1 containing all elements that map to the identity of G_2. This subset must contain at least one element because every homomorphism maps the identity of G_1 to the identity of G_2.

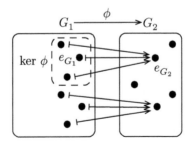

[1] There are more isomorphism theorems—investigating them would give you an idea of where Abstract Algebra goes next.

How does this apply to the homomorphisms considered above? For some, the kernel contains multiple elements of G_1. For instance, the homomorphism $\phi : \mathbb{Z}_6 \to \mathbb{Z}_6$ with $\phi(1) = 3$ has kernel $\{0, 2, 4\}$; the homomorphism $\phi : (\mathbb{R}, +) \to (U, \times)$ given by $\phi(x) = e^{2\pi i x}$ has kernel $\mathbb{Z}$. For others, the kernel contains only one element: the homomorphism $\phi : (\mathbb{Z}, +) \to (\mathbb{Q}, +)$ with $\phi(1) = 2$ has kernel $\{0\}$. Note that $\{0\}$ is the set containing the element zero, not just the number 0—you could probably get away with writing 'ker $\phi = 0$', but mathematicians notice this kind of technical inaccuracy. Note also that this third homomorphism, with only the identity in its kernel, is injective (where the others are not). This is not a coincidence, and the general result is captured below. As usual, remember the self-explanation training from Section 3.6. Can you justify each deduction in this proof with definitional properties or theorems about homomorphisms?

Theorem: Suppose that $\phi : G_1 \to G_2$ is a group homomorphism, and that $\ker \phi = \{e_{G_1}\}$. Then ϕ is injective.

Proof: Suppose that $\phi : G_1 \to G_2$ is a group homomorphism with $\ker \phi = \{e_{G_1}\}$.

Suppose that $\exists a_1, a_2 \in G_1$ such that $\phi(a_1) = \phi(a_2)$.

Then

$$\phi(a_1)(\phi(a_2))^{-1} = \phi(a_2)(\phi(a_2))^{-1}$$
$$\Rightarrow \phi(a_1)(\phi(a_2))^{-1} = e_{G_2}$$
$$\Rightarrow \phi(a_1)\phi(a_2^{-1}) = e_{G_2}$$
$$\Rightarrow \phi(a_1 a_2^{-1}) = e_{G_2}$$
$$\Rightarrow a_1 a_2^{-1} = e_{G_1} \text{ because } \ker \phi = \{e_{G_1}\}$$
$$\Rightarrow a_1 = a_2.$$

Thus ϕ is injective.

A second key result is that, regardless of how many elements it contains, the kernel of a homomorphism is always a *normal subgroup*. Normal subgroups—discussed in Section 7.5—are defined as below.

Definition: Let H be a subgroup of G. Then H is a **normal subgroup** if and only if $\forall a \in G$, $aH = Ha$.

To prove that the kernel K of a homomorphism $\phi : G_1 \to G_2$ must be a normal subgroup, we must therefore prove that it is a subgroup and that for every $a \in G_1$, $aK = Ka$. Can you see how both might follow from the property that for every $k \in K$, $\phi(k) = 0$? Think about that as you read this theorem and proof.

Theorem: Suppose that $K = \ker \phi$ where $\phi : G_1 \to G_2$ is a group homomorphism. Then K is a normal subgroup of G_1.

Proof: First we will prove that K is a subgroup of G_1.

K is closed under the operation for G_1 because for every $k_1, k_2 \in K$, $\phi(k_1 k_2) = \phi(k_1)\phi(k_2) = e_{G_2} e_{G_2} = e_{G_2}$.

The operation on K is associative because it is inherited from G_1.

The identity $e_{G_1} \in K$ because ϕ is a homomorphism so $\phi(e_{G_1}) = e_{G_2}$.

If $k \in K$ then $k^{-1} \in K$ because ϕ is a homomorphism so $\phi(k^{-1}) = (\phi(k))^{-1} = (e_{G_2})^{-1} = e_{G_2}$.

Thus K is a subgroup of G_1.

Now let $a \in G_1$.

Then $g \in aK \Leftrightarrow \exists k \in K$ such that $g = ak$.

For this k, $\phi(g) = \phi(ak) = \phi(a)\phi(k) = \phi(a)e_{G_2} = \phi(a)$.

So $aK = \{g \in G_1 | \phi(g) = \phi(a)\}$.

Similarly, $g \in Ka \Leftrightarrow \exists k \in K$ such that $g = ka$.

For this k, $\phi(g) = \phi(ka) = \phi(k)\phi(a) = e_{G_2}\phi(a) = \phi(a)$.

So $Ka = \{g \in G_1 | \phi(g) = \phi(a)\}$.

Thus $aK = Ka$, so K is a normal subgroup of G_1.

This proof establishes equality of left and right cosets of K in G_1 by showing that for a given $a \in G$, both comprise all elements $g \in G_1$ for which $\phi(g) = \phi(a)$. This formalizes the way in which a homomorphism organizes elements of its domain into sets that collapse to single elements of its image.

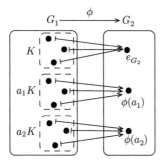

The theorem provides a new way to identify normal subgroups by identifying kernels of homomorphisms. For instance, $(\{0,2,4\}, +_6)$ is a normal subgroup of $\mathbb{Z}_6$, and $(\mathbb{Z}, +)$ is a normal subgroup of $(\mathbb{R}, +)$. We already knew that these subgroups are normal—as in Section 7.6, every subgroup of an abelian group is normal. But it is good to learn new links.

With this theory in place, we are ready for the First Isomorphism Theorem, stated below. First, read the theorem and try to understand it (this might not be easy).

First Isomorphism Theorem: Suppose that $\phi : G_1 \to G_2$ is a group homomorphism and that $K = \ker \phi$. Then $G_1/K \cong \operatorname{im} \phi$.

As ever, I like to understand abstract statements by relating them to examples. Here, for instance, the homomorphism $\phi : \mathbb{Z}_6 \to \mathbb{Z}_6$ with $\phi(1) = 3$ has $G_1 = G_2 = \mathbb{Z}_6$ and $K = \ker \phi = \{0, 2, 4\}$; the theorem concludes that $\mathbb{Z}_6/\{0, 2, 4\}$ is isomorphic to the image of ϕ, which is $(\{0, 3\}, +_6)$ (check against the previous section). This means that the cosets of $\{0, 2, 4\}$ in $\mathbb{Z}_6$ form a quotient group isomorphic to $(\{0, 3\}, +_6)$, which is also isomorphic to $\mathbb{Z}_2$.

$+_6$	0	3
0	0	3
3	3	0

$+_2$	0	1
0	0	1
1	1	0

The homomorphism $\phi : (\mathbb{R}, +) \to (U, \times)$ given by $\phi(x) = e^{2\pi i x}$ has $G_1 = (\mathbb{R}, +)$, $G_2 = (U, \times)$, and $\ker \phi = \mathbb{Z}$. The theorem concludes that $(\mathbb{R}, +)/(\mathbb{Z}, +)$ is isomorphic to the image of ϕ, which is the unit circle (every element of the unit circle is 'hit'). This means that the cosets of $(\mathbb{Z}, +)$ in $(\mathbb{R}, +)$ form a quotient group isomorphic to the unit circle under multiplication. This is a pretty profound result.

To link to ideas from earlier in the book, consider the map $\phi : (\mathbb{Z}, +) \to (\mathbb{Z}_{12}, +_{12})$ given by $\phi(x) = x \bmod 12$ (mapping x to its remainder on division by 12). This is a homomorphism because—to reformulate a result from Section 3.4—it satisfies $\phi(a + b) = \phi(a) +_{12} \phi(b)$ for every $a, b \in \mathbb{Z}$. Its kernel is

$$\ker \phi = \{x \in \mathbb{Z} | \phi(x) = 0\} = \{\ldots, -24, -12, 0, 12, 24, \ldots\} = 12\mathbb{Z},$$

and the First Isomorphism Theorem confirms that $\mathbb{Z}/12\mathbb{Z} \cong \mathbb{Z}_{12}$.

Understanding why the First Isomorphism Theorem holds is an excellent exercise in tracking object types. To this end, I think that a proof outline with a diagram is probably more illuminating than a full proof. Here is the theorem again.

First Isomorphism Theorem: Suppose that $\phi : G_1 \to G_2$ is a group homomorphism and that $K = \ker \phi$. Then: $G_1/K \cong \operatorname{im} \phi$.

The premises imply that K is a normal subgroup of G_1. So, by Section 7.5, there is a quotient group G_1/K with elements of the form aK (or Ka). By a theorem in this section, ϕ sends every element of the coset aK to $\phi(a)$, as represented on the left below. This makes it possible to think 'up a level', defining a new map Φ ('big phi') from quotient group elements to $\operatorname{im} \phi$ as represented on the right below. In notation, $\Phi : G/K \to \operatorname{im} \phi$, where $\Phi(aK) = \phi(a)$.

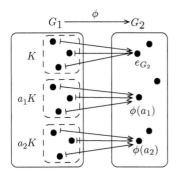 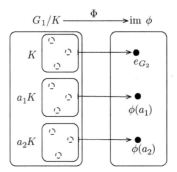

A proof of the First Isomorphism Theorem must establish that Φ is an isomorphism. Does the diagram convince you that Φ is bijective? Can you convert that understanding into a formal argument? And can you use the fact that $\Phi(aK) = \phi(a)$ to establish that for every $aK, bK \in G_1/K$, $\Phi((aK)(bK)) = \Phi(aK)\Phi(bK)$? I will leave this to your course, but you can probably work it out—if your course uses different notation then I recommend that you use it to relabel the diagram above.

To conclude this chapter, some reasoning using two major theorems: the First Isomorphism Theorem and Lagrange's Theorem (see Section 7.7). These are repeated below, with a new claim.

Lagrange's Theorem: Suppose that G is a finite group and H is a subgroup of G. Then $|G| = |H||G : H|$.

First Isomorphism Theorem: Suppose that $\phi : G_1 \to G_2$ is a group homomorphism and that $K = \ker\phi$. Then $G_1/K \cong \text{im } \phi$.

Claim: Suppose that a group G_1 has order pq where p and q are both prime, and $\phi : G_1 \to G_2$ is a surjective group homomorphism. Then $G_1 \cong G_2$ or G_2 is abelian.

To me, the claim sounds a bit odd, and I have no immediate intuition about why it should be true. But that need not bother me, because I can still try to prove it using a formal approach (see Section 3.7). The claim's premises fit with the First Isomorphism Theorem, involving a group homomorphism $\phi : G_1 \to G_2$. Moreover, ϕ is surjective, meaning that the image of ϕ is the whole of G_2, so the First Isomorphism Theorem implies that $G_1/K \cong G_2$. But what is the kernel K? It must be a subgroup of G_1 so, by Lagrange's Theorem, $|G_1| = |K||G_1{:}K|$; in words, the order of G_1 is equal to the order of K multiplied by the number of cosets of K in G_1. Specifically, the order of the kernel must divide the order of G_1, so it must be $1, p, q$ or pq. That is a manageable number of possibilities, so I can run through them to see what happens. If $|K| = 1$, then $K = \{e_{G_1}\}$ and the First Isomorphism Theorem gives $G_1 \cong G_2$. If $|K| = pq$, then K is the whole of G_1, so G_2 is a one-element group, which must be abelian. If $|K| = p$ then $G_1/K \cong G_2$ means that $|G_2| = q$, and if $|K| = q$ then $|G_2| = p$. Either way, G_2 is a group of prime order so it has no nontrivial proper subgroups and must be cyclic, and thus abelian. So my proof is complete.

As ever, I do not intend to imply that you should have such reasoning at your fingertips—understanding the above paragraph might take some rereading and reference to earlier chapters. I do intend to imply that simple actions can improve your chances of constructing proofs for new claims. Abstract Algebra, with its tightly interconnected theory, rewards students who have a bit of optimism and an up-to-date list of theorems.

CHAPTER 9

Rings

This chapter introduces rings, which are sets with two binary operations that satisfy certain axioms. It discusses example rings and simple ring theorems. It then considers extra properties shared by some but not all rings, noting which are necessary for a ring to be an integral domain or a field. It relates these properties to equation solving and to theorems about these structures. Later sections introduce subrings, ideals, quotient rings and ring homomorphisms.

9.1 What is a ring?

I f you have studied group theory, you will know that the integers $\mathbb{Z}$ form a group under addition. But, as noted in Section 6.6, they form not just this group but also a *ring* under addition and multiplication. A ring is a set with two binary operations that satisfy the definition below. A general ring is often denoted $(R, +, \cdot)$ or $(R, +, \times)$ or just R. You can replace R with $\mathbb{Z}$ in each axiom below to check that $\mathbb{Z}$ satisfies the definition of a ring under standard addition and multiplication. Note that the list requires a ring to have a multiplicative identity, which is no problem for $\mathbb{Z}$. But, as noted in Section 2.4, it is possible to define rings without this axiom then call a ring with a multiplicative identity a *ring with unity* or *ring with one*. Here, observe that the notation 'R' is obviously sensible, but can lead to errors for people accustomed to writing '$\mathbb{R}$' for the real numbers.[1] Is $\mathbb{R}$ a ring under standard addition and multiplication?

---

[1] When making notes for this chapter, I kept writing '$\mathbb{R}$' then changing it to 'R'.

Definition: A **ring** is a set R with binary operations $+$ and $\cdot$ such that:

Closure under addition $\forall a, b \in R, a + b \in R$;

Associativity of addition $\forall a, b, c \in R, (a + b) + c = a + (b + c)$;

Additive identity $\exists 0 \in R$ such that $\forall a \in R, 0 + a = a + 0 = 0$;

Additive inverses $\forall a \in R, \exists(-a) \in R$ such that
$a + (-a) = (-a) + a = 0$;

Commutativity of addition $\forall a, b \in R, a + b = b + a$;

Closure under multiplication $\forall a, b \in R, a \cdot b \in R$;

Associativity of multiplication $\forall a, b, c \in R, (a \cdot b) \cdot c = a \cdot (b \cdot c)$;

Multiplicative identity $\exists 1 \in R$ such that $\forall a \in R, 1 \cdot a = a \cdot 1 = a$;

Left distributivity $\forall a, b, c \in R, a \cdot (b + c) = a \cdot b + a \cdot c$;

Right distributivity $\forall a, b, c \in R, (a + b) \cdot c = a \cdot c + b \cdot c$.

Another thing to observe is that it is common to denote the additive identity by '0' and the multiplicative identity by '1', although rings do not have to be based on sets of numbers. A third thing to observe is that, as noted in Section 2.5, the definition looks long. But the first five axioms specify that R is an *abelian group under addition*, meaning an additive group in which the operation is commutative. So the ring definition can be abbreviated.

Definition: A **ring** is a set R with binary operations $+$ and $\cdot$ such that:

Additive abelian group R is an abelian group under $+$;

Closure under multiplication $\forall a, b \in R, a \cdot b \in R$;

Associativity of multiplication $\forall a, b, c \in R, (a \cdot b) \cdot c = a \cdot (b \cdot c)$;

Multiplicative identity $\exists 1 \in R$ such that $\forall a \in R, 1 \cdot a = a \cdot 1 = a$;

Left distributivity $\forall a, b, c \in R, a \cdot (b + c) = a \cdot b + a \cdot c$;

Right distributivity $\forall a, b, c \in R, (a + b) \cdot c = a \cdot c + b \cdot c$.

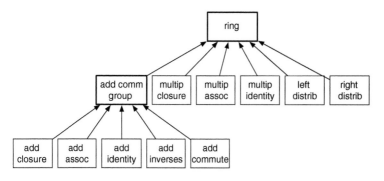

When writing the definition, lecturers might abbreviate further by omitting the axiom names. Indeed, they might omit the closure axioms altogether because closure is assumed in the definition of *binary operation* (see Section 5.8). But I like to name everything and to list closure explicitly to ensure that it is not forgotten; I also think it helps when introducing subrings, as in Section 9.6.

Notation for rings mirrors that for groups. Just as writing $(\mathbb{Z}, +)$ emphasizes the additive group structure of the integers, writing $(\mathbb{Z}, +, \cdot)$ or $(\mathbb{Z}, +, \times)$ emphasizes their two-operation ring structure. A course might not use that notation, though. As elsewhere in mathematics, multiplication is commonly denoted by juxtaposition, using $(ab)c = a(bc)$ rather than $(a \cdot b) \cdot c = a \cdot (b \cdot c)$ or $(a \times b) \times c = a \times (b \times c)$. This merits attention if you have previously studied group theory. Each group has just one operation, so group theoretic claims are often written using juxtaposition regardless of whether the operation is multiplication or addition or composition or something else. Each ring has *two* operations, which need to be distinguished. So, although ring multiplication might be denoted in different ways, addition is usually denoted explicitly by '+'. This will be important in Section 9.7.

To conclude this opening section, two anticipatory notes about the ring definition and how it is used. First, as noted in Section 2.5, the ring axioms might seem lopsided: addition has to be commutative but multiplication does not, and a ring must have additive inverses but not necessarily multiplicative inverses. A ring *can* have those extra properties—for instance, $\mathbb{Z}$ has commutative multiplication. But it doesn't have to—$\mathbb{Z}$ does not have multiplicative inverses (with exceptions for which two elements?). Extra properties are discussed in Section 9.4. Second, most ring axioms deal only with addition or only with multiplication (check). But the distributivity axioms combine addition and multiplication, meaning that distributivity is often useful in proving theorems about rings such as those to appear in Section 9.3.

9.2 Examples of rings

Various familiar sets form rings under standard addition and multiplication: $\mathbb{Z}$, $\mathbb{Q}$, $\mathbb{R}$ and $\mathbb{C}$ are all rings. You can check against the axioms, and you might be asked to do so in a course. In some cases, a lecturer might just assert that they hold—stating, for instance, that for every $a, b \in \mathbb{R}$, $a + b \in \mathbb{R}$. In other cases, manipulation is appropriate to establish that everything has the required form. For instance, the following shows that $\mathbb{Q}$ is closed under addition.

Claim: $\forall a, b \in \mathbb{Q}, a + b \in \mathbb{Q}$.

Proof: Suppose that $a, b \in \mathbb{Q}$.

Then $\exists p, r \in \mathbb{Z}$ and $q, s \in \mathbb{Z}\backslash\{0\}$ such that $a = \dfrac{p}{q}$ and $b = \dfrac{r}{s}$.

So $a + b = \dfrac{p}{q} + \dfrac{r}{s} = \dfrac{ps + rq}{qs} \in \mathbb{Q}$ because $ps + rq \in \mathbb{Z}$ and $qs \in \mathbb{Z}\backslash\{0\}$.

What would you write to show that $\mathbb{C}$ is closed under multiplication? And what else do you know about $\mathbb{Z}$, $\mathbb{Q}$, $\mathbb{R}$ and $\mathbb{C}$? What properties do they have that go beyond the ring axioms and distinguish them from one another? If you have never studied this, you might find it surprisingly hard to pin down—the next section will look at some distinctions. For now, it is worth exploring where rings occur in other familiar sets. For instance, if $n \in \mathbb{N}$, is $n\mathbb{Z} = \{nx | x \in \mathbb{Z}\}$ a ring? It has addition and multiplication, so we can start exploring the axioms. Which hold and which, if any, do not?

Closure under addition holds, but does require checking because not every subset of $\mathbb{Z}$ has this property. For instance, the set of odd numbers is not closed under addition because adding two odds gives an even. That $3\mathbb{Z}$ *is* closed under addition can be demonstrated as below. This might seem like overkill for something so simple, but you might be expected to spell out such arguments, especially in the early stages of a course when the focus is on building proofs from axioms.

Claim: $\forall a, b \in 3\mathbb{Z}, a + b \in 3\mathbb{Z}.$

Proof: Suppose that $a, b \in 3\mathbb{Z}.$

Then $\exists x, y \in \mathbb{Z}$ such that $a = 3x$ and $b = 3y$.

So $a + b = 3x + 3y = 3(x + y)$, where $x + y \in \mathbb{Z}$.

Hence $a + b \in 3\mathbb{Z}$.

Associativity of addition in $3\mathbb{Z}$ is *inherited* from $\mathbb{Z}$ because associativity is defined using a single universal quantifier (see Section 3.5): for all $a, b, c \in \mathbb{Z}$, $(a + b) + c = a + (b + c)$. And $3\mathbb{Z} \subseteq \mathbb{Z}$, so it follows that for all $a, b, c \in 3\mathbb{Z}$, $(a + b) + c = a + (b + c)$.

Next, there is an additive identity in $3\mathbb{Z}$ because $0 = 3 \times 0 \in 3\mathbb{Z}$ and for all $a \in 3\mathbb{Z}$, $0 + a = a + 0 = 0$. But attention to identities might make you notice that there is no multiplicative identity: $1 \notin 3\mathbb{Z}$ and there is no other element with the required property. So $3\mathbb{Z}$ is not a ring by our definition.[2] That is the only ring axiom not satisfied, though. You might like to consider the rest: they all hold, but which are inherited from $\mathbb{Z}$, and which warrant checks?

A different type of structure is $(\mathbb{Z}_n, +_n, \times_n)$, the integers modulo n under addition and multiplication modulo n. This is usually denoted simply '$\mathbb{Z}_n$', and its operations were first discussed in Section 3.3. For $\mathbb{Z}_7$, addition and multiplication tables appear below. Is $\mathbb{Z}_7$ a ring? If you think that it is, is that because 7 is special in some way, or is $\mathbb{Z}_n$ a ring for every natural number n?

---

[2] As noted in Section 2.4, in some areas of mathematics it makes sense to omit the multiplicative identity axiom. In that case, $3\mathbb{Z}$ does qualify as a ring, just not a *ring with unity* or *ring with one*.

$+_7$	0	1	2	3	4	5	6
0	0	1	2	3	4	5	6
1	1	2	3	4	5	6	0
2	2	3	4	5	6	0	1
3	3	4	5	6	0	1	2
4	4	5	6	0	1	2	3
5	5	6	0	1	2	3	4
6	6	0	1	2	3	4	5

$\times_7$	0	1	2	3	4	5	6
0	0	0	0	0	0	0	0
1	0	1	2	3	4	5	6
2	0	2	4	6	1	3	5
3	0	3	6	2	5	1	4
4	0	4	1	5	2	6	3
5	0	5	3	1	6	4	2
6	0	6	5	4	3	2	1

It turns out that $\mathbb{Z}_7$ is a ring and every $\mathbb{Z}_n$ is a ring. But the tables support thinking only to a limited extent. We can 'see' closure under addition and multiplication because everything in both tables is an element of $\mathbb{Z}_7$. In the left table we can 'see' that the additive identity is 0, that each element has an additive inverse, and that addition is commutative (review Section 5.1 if you are not sure how). But this 'seeing' is informal. What would you write to explain more fully?

For other axioms, we run into trouble. As noted in Section 5.3, tables show the results of combining any two elements—they represent *binary* operations. For axioms involving three elements, they haven't enough dimensions. Moreover, properties of $\mathbb{Z}_n$ cannot be inherited directly from $\mathbb{Z}$ because the elements and operations are different. Fortunately, both addition and multiplication modulo n work so that all of the axioms are satisfied. For instance, for left distributivity, the remainder of $a(b + c)$ on division by n is always the same as the remainder of $ab + ac$ on division by n. Why? How do such properties in $\mathbb{Z}_n$ rely indirectly on properties in $\mathbb{Z}$?

A course might explore this in more or less detail, establishing or just stating that $\mathbb{Z}$, $\mathbb{Q}$, $\mathbb{R}$ and $\mathbb{C}$ are rings, and that if n is a natural number then so is $\mathbb{Z}_n$. Have we encountered other rings? Some groups studied in this book do not have a second binary operation: symmetries, for instance, are combined only using composition. But some do: square matrices have both addition and multiplication. For 2×2 matrices, these are defined as below.

$$\begin{pmatrix} a & b \\ c & d \end{pmatrix} + \begin{pmatrix} w & x \\ y & z \end{pmatrix} = \begin{pmatrix} a+w & b+x \\ c+y & d+z \end{pmatrix};$$

$$\begin{pmatrix} a & b \\ c & d \end{pmatrix} \begin{pmatrix} w & x \\ y & z \end{pmatrix} = \begin{pmatrix} aw+by & ax+bz \\ cw+dy & cx+dz \end{pmatrix}.$$

The set of all 2×2 matrices with entries in $\mathbb{R}$ can be denoted by $M_{2\times 2}(\mathbb{R})$. Is $M_{2\times 2}(\mathbb{R})$ a ring? It is. Some axioms are easy to check: closure is straightforward because adding or multiplying two 2×2 matrices gives another (note that $aw + by$ is a single number). The additive identity is the matrix of zeros, and

$$\begin{pmatrix} a & b \\ c & d \end{pmatrix} \text{ has additive inverse } \begin{pmatrix} -a & -b \\ -c & -d \end{pmatrix}.$$

This holds because 0 is the additive identity in $\mathbb{R}$ and $a, b, c, d \in \mathbb{R}$ have additive inverses $-a, -b, -c, -d \in \mathbb{R}$: the ring properties in $M_{2\times 2}(\mathbb{R})$ rely on ring properties in $\mathbb{R}$. Similarly, commutativity of addition in $M_{2\times 2}(\mathbb{R})$ relies on commutativity of addition in $\mathbb{R}$. At which step below is that property used? And what would you write to demonstrate that associativity and distributivity hold in $M_{2\times 2}(\mathbb{R})$? Maybe work through the calculations.

$$\begin{aligned} \begin{pmatrix} a & b \\ c & d \end{pmatrix} + \begin{pmatrix} w & x \\ y & z \end{pmatrix} &= \begin{pmatrix} a+w & b+x \\ c+y & d+z \end{pmatrix} \\ &= \begin{pmatrix} w+a & x+b \\ y+c & z+d \end{pmatrix} \\ &= \begin{pmatrix} w & x \\ y & z \end{pmatrix} + \begin{pmatrix} a & b \\ c & d \end{pmatrix}. \end{aligned}$$

A course might prove that because $\mathbb{R}$ satisfies each ring axiom, so does $M_{2\times 2}(\mathbb{R})$. Or it might do something more general, proving that if R is a ring then so is $M_{2\times 2}(R)$, and thereby establishing that $M_{2\times 2}(\mathbb{Z})$, $M_{2\times 2}(\mathbb{Q})$, $M_{2\times 2}(\mathbb{R})$ and $M_{2\times 2}(\mathbb{C})$ are all rings. Indeed, if R is a ring and n

is a natural number then $M_{n \times n}(R)$ is a ring. I would not want to check that for $n \geq 3$, but if you do then that can only consolidate your understanding.

A final type of ring often studied is a *ring of polynomials*. For example, $\mathbb{Z}[x]$ denotes the *ring of polynomials with integer coefficients*, meaning all expressions of the form

$$a_n x^n + a_{n-1} x^{n-1} + \cdots + a_2 x^2 + a_1 x + a_0,$$

where n is a non-negative integer and the coefficients a_i are in $\mathbb{Z}$. The polynomials below are all elements of $\mathbb{Z}[x]$.

$$x^2 + 3x - 1 \qquad x - 4 \qquad -x^{10} - 59x^3 \qquad 5$$

However, the expression '5' might feel like a number, not a polynomial, or indeed a poly-anything. It is a number, of course, but it is also a polynomial—it just happens to have $a_i = 0$ for every $i \neq 0$. You might instinctively dislike that, but for $\mathbb{Z}[x]$ to be a ring, it *must* include 5. Otherwise, what would happen for this polynomial sum?

$$(x^2 - 3x + 5) + (-x^2 + 3x)$$

If 5 were not a polynomial, then the result would not be in $\mathbb{Z}[x]$, so $\mathbb{Z}[x]$ would not be closed under addition. That is surely more untidiness than we would want. A similar observation might clear up any hesitancy about 5 as a complex number: if it were not, then $(7 + i) + (-2 - i)$ would not be in $\mathbb{C}$, so $\mathbb{C}$ would not be closed under addition. When people say that Abstract Algebra provides a more sophisticated view of earlier mathematics, this is the kind of thing they mean.

Now, I just asked you to consider a sum of polynomials without explaining how to add in $\mathbb{Z}[x]$. That is because you already know. Addition works like this.

$$(a_n x^n + a_{n-1} x^{n-1} + \cdots + a_1 x + a_0) + (b_n x^n + b_{n-1} x^{n-1} + \cdots + b_1 x + b_\bullet$$
$$= (a_n + b_n)x^n + (a_{n-1} + b_{n-1})x^{n-1} + \cdots + (a_1 + b_1)x + (a_0 + b_0).$$

The general formulation can make it seem that the two polynomials must have the same degree. But some coefficients could be zero—you could find this sum:

$$(2x^3 + 6x - 10) + (3x^2 - 3x + 29).$$

Polynomial multiplication is harder. It can be expressed in double sigma notation, which is compact but sometimes pointlessly so because many people need to unpack it to understand what it means.

$$\sum_{k=0}^{n} a_k x^k \sum_{k=0}^{n} b_k x^k = \sum_{k=0}^{2n} \sum_{i=0}^{k} (a_i b_{k-i}) x^k$$

I find it easier to reorder so that low powers appear first. Below, what would be some of the intermediate terms and how do they relate to the sigma notation?

$$(a_0 + a_1 x + a_2 x^2 + \cdots + a_n x^n)(b_0 + b_1 x + b_2 x^2 + \cdots + b_n x^n)$$
$$= (a_0 b_0) + (a_0 b_1 + a_1 b_0)x + (a_0 b_2 + a_1 b_1 + a_2 b_0)x^2 + \cdots + a_n b_n x^{2n}.$$

$\times$	a_0	$a_1 x$	$a_2 x^2$	$\cdots$	$a_n x^n$
b_0	$a_0 b_0$	$a_1 b_0 x$	$a_2 b_0 x^2$	$\cdots$	
$b_1 x$	$a_0 b_1 x$	$a_1 b_1 x^2$			
$b_2 x^2$	$a_0 b_2 x^2$				
$\vdots$					$\vdots$
$b_n x^n$				$\cdots$	$a_n b_n x^{2n}$

Now, I said that $\mathbb{Z}[x]$ is a ring. Are you convinced that it satisfies the axioms? Below, these appear with $R = \mathbb{Z}[x]$ and with a, b, c replaced by $p(x), q(x), r(x)$ so that the elements look like polynomials (which is not necessary but can counteract the tendency to think of everything as a number). Do some axioms hold more obviously than others? Although none are inherited from $\mathbb{Z}$, do some follow as direct consequences of the fact that the coefficients are in $\mathbb{Z}$? Are $\mathbb{Q}[x], \mathbb{R}[x]$ and $\mathbb{C}[x]$ rings too?

Closure under addition $\forall p(x), q(x) \in \mathbb{Z}[x], p(x) + q(x) \in \mathbb{Z}[x]$;

Associativity of addition $\forall p(x), q(x), r(x) \in \mathbb{Z}[x],$
$(p(x) + q(x)) + r(x) = p(x) + (q(x) + r(x))$;

Additive identity $\exists 0 \in \mathbb{Z}[x]$ such that $\forall p(x) \in \mathbb{Z}[x],$
$0 + p(x) = p(x) + 0 = 0$;

Additive inverses $\forall p(x) \in \mathbb{Z}[x], \exists (-p(x)) \in \mathbb{Z}[x]$ such that
$p(x) + (-p(x)) = (-p(x)) + p(x) = 0$;

Commutativity of addition $\forall p(x), q(x) \in \mathbb{Z}[x],$
$p(x) + q(x) = q(x) + p(x)$;

Closure under multiplication $\forall p(x), q(x) \in \mathbb{Z}[x], p(x)q(x) \in \mathbb{Z}[x]$;

Associativity of multiplication $\forall p(x), q(x), r(x) \in \mathbb{Z}[x],$
$(p(x)q(x))r(x) = p(x)(q(x)r(x))$;

Multiplicative identity $\exists 1 \in \mathbb{Z}[x]$ such that $\forall p(x) \in \mathbb{Z}[x],$
$1p(x) = p(x)1 = p(x)$;

Left distributivity $\forall p(x), q(x), r(x) \in \mathbb{Z}[x],$
$p(x)(q(x) + r(x)) = p(x)q(x) + p(x)r(x)$;

Right distributivity $\forall p(x), q(x), r(x) \in \mathbb{Z}[x],$
$(p(x) + q(x))r(x) = p(x)q(x) + p(x)r(x)$.

Finally, are the ring axioms now familiar enough that, if you closed this book, you could write them out? Try it—this would be good retrieval practice.

9.3 Simple ring theorems

Rings of numbers, matrices and polynomials satisfy the ring definition, so theorems deduced from its axioms apply to them all. Below are two simple theorems often encountered early in ring theory.

Theorem: Let R be a ring. Then for every $a \in R$, $0 \cdot a = a \cdot 0 = 0$.

Theorem: Let R be a ring. Then for every $a, b \in R$,

$$(-a)b = a(-b) = -(ab) \text{ and } (-a)(-b) = ab.$$

These theorems look a lot like the axioms—they are no more complicated to state. They have the status of theorems, though, because they can be proved from the axioms. I will provide proofs, but first we will consider why these theorems are worth stating and how they relate to rings.

The first theorem probably seems obvious. You have known for a decade that 'anything times zero is zero', so you might wonder why mathematicians would bother proving it now. But why exactly does it hold for the additive identity in every ring, not just in rings of numbers? Even stating this theorem demands something like a ring structure because it involves an additive identity with a multiplicative operation. This might remind you that two ring axioms link addition and multiplication—which two? Read the proof below to work out which are used where.

Theorem: Let R be a ring. Then for every $a \in R$, $0 \cdot a = a \cdot 0 = 0$.

Proof: Let $a \in R$.

Note that 0 is the additive identity in R so $0 + 0 = 0$.

Hence $(0 + 0)a = 0a$.

So $0a + 0a = 0a$.

Now $0a \in R$ so $0a$ has an additive inverse $-0a \in R$.

Thus
$$0a + 0a = 0a$$
$$\Rightarrow 0a + 0a + (-0a) = 0a + (-0a)$$
$$\Rightarrow \qquad\qquad 0a = 0.$$

By a similar argument, $a0 = 0$.

Then ask, why is each step is valid? Would you have thought to use $(0 + 0)a = 0a$? Either way, can you see that it is a clever move? And can you write out the 'similar argument'?

The second theorem—repeated below—might seem less obvious.

Theorem: Let R be a ring. Then for every $a, b \in R$,
$$(-a)b = a(-b) = -(ab) \text{ and } (-a)(-b) = ab.$$

When you first met the claim that $(-a)(-b) = ab$ for numbers, it might have seemed arbitrary or mysterious; you might have been unsatisfied with claims that 'a minus times a minus is a plus'. If so, good for you—a mathematically minded person should always look for justification. I, for instance, reasoned that $3 \cdot (-2)$ was three 'lots of' -2, and $(-3) \cdot (-2)$ must be 'minus' that. The proof below formalizes that reasoning. It demonstrates that the interaction between additive inverses and multiplication is not arbitrary or mysterious but a necessary consequence of the ring axioms.

Theorem: Let R be a ring. Then for every $a, b \in R$,

$$(-a)b = a(-b) = -(ab) \text{ and } (-a)(-b) = ab.$$

Proof: Let $a, b \in R$.

Then a has an additive inverse $(-a) \in R$ with $a + (-a) = 0$.

So $(a + (-a))b = 0b = 0$.

Thus $ab + (-a)b = 0$.

Now $ab \in R$ so ab has an additive inverse $-(ab) \in R$.

Thus
$$ab + (-a)b = 0$$
$$\Rightarrow -(ab) + ab + (-a)b = -(ab) + 0$$
$$\Rightarrow (-a)b = -(ab).$$

By a similar argument, $a(-b) = -(ab)$.

Next note that b has an additive inverse $(-b) \in R$ with $b + (-b) = 0$.

So
$$(-a)(b + (-b)) = 0$$
$$\Rightarrow (-a)b + (-a)(-b) = 0$$
$$\Rightarrow -(ab) + (-a)(-b) = 0 \qquad \text{by the above}$$
$$\Rightarrow ab + (-(ab)) + (-a)(-b) = ab + 0$$
$$\Rightarrow (-a)(-b) = ab.$$

Again, why is each step valid? If your lecturer were fussy about including axiomatic justifications—'by distributivity' and so on—what might you write? Would you have thought to use $(a + (-a))b = 0b = 0$? Might you think of it in future?

To conclude, a note on proofs like these. When I studied ring theory, I thought it ingenious to add two zeros or an element and its inverse, then split the expression to derive the desired result. But I don't think I noticed that such arguments rely on distributivity. I didn't notice that this had to be the case for theorems linking additive identities or inverses to a multiplicative operation. I certainly didn't grasp the fact that two operations and a ring structure were necessary for such results to be meaningful. If I had noticed all of that, I think I would have found theorems and proofs like these easier to reconstruct and more satisfying to study.

9.4 Rings, integral domains and fields

The previous section's proofs use only the ring axioms, so the corresponding theorems apply to every ring. But, as noted earlier, extra properties can be appended to the ring axioms to define other structures. For instance, the ring $\mathbb{Z}$ has commutative multiplication, which makes it a *commutative ring*.

Definition: A ring R is **commutative** if and only if $\forall a, b \in R$, $ab = ba$.

Note that all rings have commutative addition; only those that also have commutative multiplication are called commutative rings. This means that the matrix ring $M_{2 \times 2}(\mathbb{R})$ is not a commutative ring. The Venn diagram below captures the fact that commutative rings form a subset of all rings. Check that each example is correctly placed (the spacing is weird for a reason—bear with me).

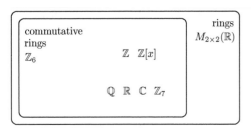

What $\mathbb{Z}$ does not have is multiplicative inverses: it is not true that for all $a \in \mathbb{Z}$, there exists $a^{-1} \in \mathbb{Z}$ such that $aa^{-1} = a^{-1}a = 1$. This means that $\mathbb{Z}$ is not a *division ring*.

Definition: A ring R is a **division ring** if and only if $\forall a \in R \setminus \{0\}\ \exists a^{-1} \in R$ such that $aa^{-1} = a^{-1}a = 1$.

The matrix ring $M_{2 \times 2}(\mathbb{R})$ is not a division ring either: only matrices with nonzero determinants are *invertible*. And the ring of polynomials $\mathbb{Z}[x]$ is not a division ring—which of its elements do and do not have multiplicative inverses? But $\mathbb{Q}$, $\mathbb{R}$ and $\mathbb{C}$ all satisfy the definition. And something interesting happens for rings of the form $\mathbb{Z}_n$. In some, every element except 0 has a multiplicative inverse. In others, that is not the case. What is the distinction, and how does it play out in the multiplication tables for $\mathbb{Z}_7$ and $\mathbb{Z}_6$?

$\times_7$	0	1	2	3	4	5	6
0	0	0	0	0	0	0	0
1	0	1	2	3	4	5	6
2	0	2	4	6	1	3	5
3	0	3	6	2	5	1	4
4	0	4	1	5	2	6	3
5	0	5	3	1	6	4	2
6	0	6	5	4	3	2	1

$\times_6$	0	1	2	3	4	5
0	0	0	0	0	0	0
1	0	1	2	3	4	5
2	0	2	4	0	2	4
3	0	3	0	3	0	3
4	0	4	2	0	4	2
5	0	5	4	3	2	1

In $\mathbb{Z}_7$, every nonzero element has a multiplicative inverse:

$$1 \times_7 1 = 1, \quad 6 \times_7 6 = 1, \quad 2 \times_7 4 = 4 \times_7 2 = 1, \quad 3 \times_7 5 = 5 \times_7 3 = 1.$$

In $\mathbb{Z}_6$, the elements 2, 3 and 4 do not have multiplicative inverses—the multiplicative identity appears nowhere in their rows. This is because 2, 3 and 4 share factors with 6; they are not *relatively prime* to the n in $\mathbb{Z}_n$. In general, $\mathbb{Z}_p$ for p prime is a division ring, and $\mathbb{Z}_n$ for n composite is not. Check that everything is in the right place in this new diagram.

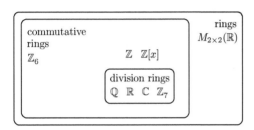

What stops $\mathbb{Z}_6$ being a division ring is that 2, 3 and 4 are all *zero divisors*.

Definition: In a ring R, a nonzero $a \in R$ is a **zero divisor** if and only if $\exists b \in R \backslash \{0\}$ such that $ab = 0$ or $ba = 0$.

Note that in some of these definitions, 0 is excluded from a quantified condition: the additive identity is 'special' in relation to some multiplicative properties. And think about other rings—which have zero divisors and which do not? How do your answers relate to the definition below?

Definition: An **integral domain** is a commutative ring with no zero divisors.

Why might such a structure be called an *integral* domain, do you think? The word 'integral' relates to *integers*, which form an integral domain— check that they have the required properties. Are any other rings discussed above integral domains? Check the placements in the diagram below.

Finally, some rings are *fields*.

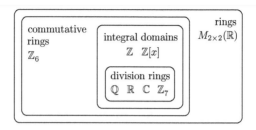

Definition: A field is a commutative division ring.

This definition is nice and short, but only because it hides numerous axioms in the words 'commutative division ring'. With the axioms spelled out, it looks like this.

Definition: A field is a set F with binary operations $+$ and $\cdot$ such that:

Closure under addition $\forall a, b \in F, a + b \in F$;

Associativity of addition $\forall a, b, c \in F, (a + b) + c = a + (b + c)$;

Additive identity $\exists 0 \in F$ such that $\forall a \in F, 0 + a = a + 0 = 0$;

Additive inverses $\forall a \in F, \exists (-a) \in F$ such that
  $a + (-a) = (-a) + a = 0$;

Commutativity of addition $\forall a, b \in F, a + b = b + a$;

Closure under multiplication $\forall a, b \in F, a \cdot b \in F$;

Associativity of multiplication $\forall a, b, c \in F, (a \cdot b) \cdot c = a \cdot (b \cdot c)$;

Multiplicative identity $\exists 1 \in F$ such that $\forall a \in F, 1 \cdot a = a \cdot 1 = a$;

Multiplicative inverses $\forall a \in F \backslash \{0\}, \exists a^{-1} \in F$ such that
  $a a^{-1} = a^{-1} a = 1$;

Commutativity of multiplication $\forall a, b \in F, a \cdot b = b \cdot a$;

Left distributivity $\forall a, b, c \in F, a \cdot (b + c) = a \cdot b + a \cdot c$;

Right distributivity $\forall a, b, c \in F, (a + b) \cdot c = a \cdot c + b \cdot c$.

Because fields must satisfy more axioms, there are fewer fields than rings. But $\mathbb{Q}$, $\mathbb{R}$ and $\mathbb{C}$ are all fields—check that you believe this. For me, the word 'field' thus seems appropriate because it conjures up an image of a big thing with lots of things in it (blades of grass, maybe). But in fact fields need not be big: $\mathbb{Z}_7$ has only seven elements, but is a field because it is commutative and has multiplicative inverses. There do exist division rings that are not fields—one example is the *quaternions*, which I will not introduce here but which you might want to look up. Because all of our listed division rings are fields, we can amend the diagram as below.

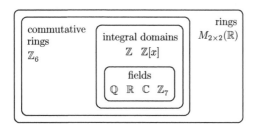

To conclude, note that every $\mathbb{Z}_p$ with p prime is a field: $\mathbb{Z}_{19}$ is a field, $\mathbb{Z}_{43}$ is a field, and so on. In particular, $\mathbb{Z}_2$ is a field. The tables below capture its structure.

$+_2$	0	1
0	0	1
1	1	0

$\times_2$	0	1
0	0	0
1	0	1

Now, a two-element field is not very interesting. But it is interesting that it exists: it is really very small. And the obvious question is, is it the smallest? If you have studied small groups—see Section 6.11—you might know that it is possible to have a one-element group. Is it possible to have a one-element field? How about a one-element ring?

9.5 Units, zero divisors and equations

Do you now understand why the ring axioms are lopsided? Fields are 'nicer' in that both operations have full sets of inverses, but rings of numbers, matrices and polynomials still have much common structure, so theory about them can be built simultaneously. They do, however, have different properties, and this section will relate these to equation solving and to the zero product property from Section 3.2.

First, a ring without a full set of multiplicative inverses might still have some. In $\mathbb{Z}$, the elements 1 and -1 have multiplicative inverses; in $\mathbb{Z}_6$, 1 and 5 have multiplicative inverses. Ring elements with multiplicative inverses are called *units*.

Definition: Suppose that R is a ring. Then $u \in R$ is a **unit** if and only if $\exists u^{-1} \in R$ such that $uu^{-1} = u^{-1}u = 1$.

Note that the multiplicative identity is always a unit: $1 \cdot 1 = 1$ must hold. Thus every ring has at least one unit. And plenty have more than one. In the multiplication table for $\mathbb{Z}_{12}$ below, rows and columns for zero divisors are in grey so that units stand out in white.

$\times_{12}$	0	1	2	3	4	5	6	7	8	9	10	11
0	0	0	0	0	0	0	0	0	0	0	0	0
1	0	1	2	3	4	5	6	7	8	9	10	11
2	0	2	4	6	8	10	0	2	4	6	8	10
3	0	3	6	9	0	3	6	9	0	3	6	9
4	0	4	8	0	4	8	0	4	8	0	4	8
5	0	5	10	3	8	1	6	11	4	9	2	7
6	0	6	0	6	0	6	0	6	0	6	0	6
7	0	7	2	9	4	11	6	1	8	3	10	5
8	0	8	4	0	8	4	0	8	4	0	8	4
9	0	9	6	3	0	9	6	3	0	9	6	3
10	0	10	8	6	4	2	0	10	8	6	4	2
11	0	11	10	9	8	7	6	5	4	3	2	1

In $M_{2\times2}(\mathbb{R})$ under multiplication, every matrix with nonzero determinant is a unit because it is *invertible*: provided $ad \neq bc$,

$$\begin{pmatrix} a & b \\ c & d \end{pmatrix} \text{ has multiplicative inverse } \frac{1}{ad-bc}\begin{pmatrix} d & -b \\ -c & a \end{pmatrix}.$$

These units form the *general linear group* of degree 2, denoted $GL(2,\mathbb{R})$. Note that $GL(2,\mathbb{R})$ is a group because multiplying together two invertible matrices gives another, associativity is inherited, the multiplicative identity matrix is invertible, and every invertible matrix by definition has a multiplicative inverse. Moreover, this generalizes: for any ring, the units form a multiplicative group. The units of $\mathbb{Z}_{12}$ are shown in the table below; they form a group isomorphic to the Klein four-group V (see Section 6.8).

$\times_{12}$	1	5	7	11
1	1	5	7	11
5	5	1	11	7
7	7	11	1	5
11	11	7	5	1

The general claim is captured below.

Claim: The set of units U in a ring R forms a multiplicative group because:

Closure $\forall u_1, u_2 \in U, u_1 u_2 \in U$;

Associativity $\forall u_1, u_2, u_3 \in U, (u_1 u_2)u_3 = u_1(u_2 u_3)$;

Identity $\exists 1 \in U$ such that $\forall u \in U, 1u = u1 = u$;

Inverses $\forall u \in U, \exists u^{-1} \in U$ such that $uu^{-1} = u^{-1}u = 1$.

In my view, two of the subclaims are easier to think about and two are harder. Do you agree? Associativity is easier because it is inherited from R. And the identity is easier because 1 is a unit. Closure is harder. Why must the product of two units be another? If u_1 has inverse u_1^{-1} and u_2 has inverse u_2^{-1}, what can we say about $u_1 u_2$? Be careful—the inverse of $u_1 u_2$ is not (necessarily) $u_1^{-1} u_2^{-1}$ because, in a ring, multiplication is not necessarily commutative. But the inverse of $u_1 u_2$ is $u_2^{-1} u_1^{-1}$, because

$$
\begin{aligned}
(u_1 u_2)(u_2^{-1} u_1^{-1}) &= u_1(u_2 u_2^{-1})u_1^{-1} \quad \text{by associativity} \\
&= u_1 u_1^{-1} \quad\quad\quad \text{because } u_2 u_2^{-1} = 1 \\
&= 1 \quad\quad\quad\quad\quad \text{because } u_1 u_1^{-1} = 1.
\end{aligned}
$$

Inverses are also harder. If u is a unit, why must its inverse u^{-1} be a unit? This might seem obvious because, in the equation $uu^{-1} = u^{-1}u = 1$, we could 'swap around' u and u^{-1} without changing anything important. Unfortunately, we do not really write about 'swapping things around' in proofs, so what can we say that is more formal? You might see an argument like the one below, which probably adds nothing to your sense of what is going on, but which handles everything politely.

Claim: Let U be the set of units in a ring R with unity.

  Then $\forall u \in U, \exists u^{-1} \in U$ such that $uu^{-1} = u^{-1}u = 1$.

Proof: Let $u \in U$, so $\exists u^{-1} \in R$ such that $uu^{-1} = u^{-1}u = 1$.

  Then $u^{-1}u = uu^{-1} = 1$, i.e. u^{-1} has inverse $u \in R$.

  Hence $u^{-1} \in U$.

These arguments together establish that the units in a ring R form a group U under the ring's multiplicative operation. But notice that U might not be a subgroup of R, because R might not be a multiplicative group. Maybe pause to think about that.

Now, the fact that the units form a multiplicative group relates closely to equation solving. In any group $(G, *)$, every element g has an inverse under the single operation $*$. So cancellation can be performed for every group element:

$$g * x = g * y \implies g^{-1} * g * x = g^{-1} * g * y \implies x = y.$$

In a ring, cancellation via multiplicative inverses works only for units. In $\mathbb{Z}_6$, for instance, the element 5 is a unit and $5^{-1} = 5$, so

$$5x = 5y \implies 5^{-1}5x = 5^{-1}5y \implies x = y.$$

Also, in the multiplication table for $\mathbb{Z}_6$, the row for 5 contains every element exactly once. Thus every element is $5 \times_6 \textit{something}$, and the equation $5x = b$ has a unique solution for every $b \in \mathbb{Z}_6$.

$\times_6$	0	1	2	3	4	5
0	0	0	0	0	0	0
1	0	1	2	3	4	5
2	0	2	4	0	2	4
3	0	3	0	3	0	3
4	0	4	2	0	4	2
5	0	5	4	3	2	1

In contrast, 2 is not a unit, so $2x = 2y$ does not imply that $x = y$. For instance, $2 \times_6 2 = 2 \times_6 5$, but this does not imply that $2 = 5$. And the row for 2 does not contain every element exactly once: $0, 2$ and 4 appear 'too many times' and $1, 3$ and 5 do not appear at all. This means that not every element is $2 \times_6 \textit{something}$: $2x = 0, 2x = 2$ and $2x = 4$ have multiple solutions, but $2x = 1, 2x = 3$ and $2x = 5$ have no solutions.

Now, in the ring $\mathbb{Z}_6$, every nonzero element is either a unit or a zero divisor (check). But how are units and zero divisors linked in general? Can ring elements be neither units nor zero divisors? Or both units and zero divisors? What do you think? It might help to explore other structures with zero divisors, such as matrix rings. In $M_{2\times2}(\mathbb{R})$, many elements are zero divisors and many are units. For instance,

$$\begin{pmatrix} 0 & 1 \\ 0 & 0 \end{pmatrix} \text{ and } \begin{pmatrix} 0 & 2 \\ 0 & 0 \end{pmatrix}$$

are both zero divisors because

$$\begin{pmatrix} 0 & 1 \\ 0 & 0 \end{pmatrix} \begin{pmatrix} 0 & 2 \\ 0 & 0 \end{pmatrix} = \begin{pmatrix} 0+0 & 0+0 \\ 0+0 & 0+0 \end{pmatrix} = \begin{pmatrix} 0 & 0 \\ 0 & 0 \end{pmatrix}.$$

And

$$\begin{pmatrix} 1 & 0 \\ 0 & 2 \end{pmatrix} \text{ and } \begin{pmatrix} 1 & 0 \\ 0 & \frac{1}{2} \end{pmatrix}$$

are both units because

$$\begin{pmatrix} 1 & 0 \\ 0 & 2 \end{pmatrix} \begin{pmatrix} 1 & 0 \\ 0 & \frac{1}{2} \end{pmatrix} = \begin{pmatrix} 1+0 & 0+0 \\ 0+0 & 0+1 \end{pmatrix} = \begin{pmatrix} 1 & 0 \\ 0 & 1 \end{pmatrix}$$

and

$$\begin{pmatrix} 1 & 0 \\ 0 & \frac{1}{2} \end{pmatrix} \begin{pmatrix} 1 & 0 \\ 0 & 2 \end{pmatrix} = \begin{pmatrix} 1+0 & 0+0 \\ 0+0 & 0+1 \end{pmatrix} = \begin{pmatrix} 1 & 0 \\ 0 & 1 \end{pmatrix}.$$

But is every element of $M_{2 \times 2}(\mathbb{R})$ either a zero divisor or a unit? Or is there 'room' in this bigger ring for elements that are neither or both? I recommend exploring.

In general, it turns out that no ring element can be both a unit and a zero divisor because zero divisors cannot be units, as proved below.

Theorem: Suppose that R is a ring and $a \in R$ is a zero divisor.

Then a is not a unit.

Proof: Suppose that $a \in R$ is a zero divisor.

Then $\exists b \in R$ with $b \neq 0$ such that $ab = 0$ or $ba = 0$.

Now suppose for contradiction that a is a unit.

Then $\exists a^{-1} \in R$ such that $a^{-1}a = aa^{-1} = 1$.

But then $ab = 0 \Rightarrow a^{-1}ab = a^{-1}0 \Rightarrow b = 0$,

and $\quad ba = 0 \Rightarrow baa^{-1} = 0a^{-1} \Rightarrow b = 0$,

either of which contradicts the assumption that $b \neq 0$.

Thus a is not a unit.

Can an element be neither unit nor zero divisor? Yes: some rings contain such elements. Did exploring in $M_{2\times 2}(\mathbb{R})$ provide any examples? The ring $\mathbb{Z}$ provides lots. Its only units are 1 and -1; almost every integer is a non-unit. But $\mathbb{Z}$ contains no zero divisors, which is part of what characterizes it as an *integral domain*.

Definition: An **integral domain** is a commutative ring with no zero divisors.

In contrast, the rings $\mathbb{Z}_6$ and $M_{2\times 2}(\mathbb{R})$ are not integral domains. And that makes them different from earlier experience. Nearly all rings in earlier algebra are integral domains, including the real numbers, the rational numbers, and the integers (all with standard operations). Because these have no zero divisors, they satisfy the *zero product property*: if $ab = 0$ then $a = 0$ or $b = 0$. This means that in integral domains, equations of the form $ax = ay$ can be solved when $a \neq 0$, *even when a has no multiplicative inverse*. For instance, 2 has no multiplicative inverse in $\mathbb{Z}$, but

$$
\begin{aligned}
2x = 2y &\Rightarrow 2x - 2y = 0 \\
&\Rightarrow 2x + 2(-y) = 0 \\
&\Rightarrow 2(x + (-y)) = 0 \\
&\Rightarrow x + (-y) = 0 \quad \text{by the zero product property} \\
&\Rightarrow x = y.
\end{aligned}
$$

Which ring and integral domain properties are used in the general argument below?

Theorem: Suppose that R is an integral domain, $a \in R$, $a \neq 0$ and $ax = ay$. Then $x = y$.

Proof:
$$
\begin{aligned}
ax = ay &\Rightarrow ax - ay = 0 \\
&\Rightarrow ax + a(-y) = 0 \\
&\Rightarrow a(x + (-y)) = 0 \\
&\Rightarrow x + (-y) = 0 \\
&\Rightarrow x = y.
\end{aligned}
$$

9.6 Subrings and ideals

This section moves on from properties of individual ring elements to *subrings*, where a subring is a subset of a ring that is a ring in its own right. For instance, $\mathbb{Z}$ is a subring of $\mathbb{Q}$. And $\mathbb{Q}$ is a subring of $\mathbb{R}$, which is a subring of $\mathbb{C}$. Similar subring relationships exist in related structures. For matrix rings, $M_{2\times2}(\mathbb{Z})$ is a subring of $M_{2\times2}(\mathbb{Q})$, which is a subring of $M_{2\times2}(\mathbb{R})$, which is a subring of $M_{2\times2}(\mathbb{C})$. For polynomial rings, $\mathbb{Z}[x]$ is a subring of $\mathbb{Q}[x]$, which is a subring of $\mathbb{R}[x]$, which is a subring of $\mathbb{C}[x]$.

Is $\mathbb{Z}_3$ a subring of $\mathbb{Z}$? No: it is not even a subset of $\mathbb{Z}$ because its elements are different. Similarly, $\mathbb{Z}_3$ is not a subring of $\mathbb{Z}_6$. Could $\mathbb{Z}_3$ be *isomorphic* to a subring of $\mathbb{Z}_6$, do you think? Does $\mathbb{Z}_6$ contain a 'copy' of $\mathbb{Z}_3$, perhaps disguised by different element names? The *group* $(\mathbb{Z}_3, +_3)$ is isomorphic to the subgroup $(\{0,2,4\}, +_6)$ of $(\mathbb{Z}_6, +_6)$, as captured in the addition tables on the left below. Can you 'see' the group isomorphism? But a ring isomorphism would require multiplicative matching too, and the multiplicative tables on the right show that $(\mathbb{Z}_3, +_3, \times_3)$ is not isomorphic to $(\{0,2,4\}, +_6, \times_6)$. Could $\mathbb{Z}_3$ be isomorphic to some other subring of $\mathbb{Z}_6$? If so, which one? If not, why not?

$+_3$	0	1	2
0	0	1	2
1	1	2	0
2	2	0	1

$\times_3$	0	1	2
0	0	0	0
1	0	1	2
2	0	2	1

$+_6$	0	1	2	3	4	5
0	0	1	2	3	4	5
1	1	2	3	4	5	0
2	2	3	4	5	0	1
3	3	4	5	0	1	2
4	4	5	0	1	2	3
5	5	0	1	2	3	4

$\times_6$	0	1	2	3	4	5
0	0	0	0	0	0	0
1	0	1	2	3	4	5
2	0	2	4	0	2	4
3	0	3	0	3	0	3
4	0	4	2	0	4	2
5	0	5	4	3	2	1

Mathematicians, of course, are interested not just in specific rings and subrings but in general theory. If R is a ring, under what conditions is S a subring? Clearly S must be a subset of R. And it must satisfy all the ring axioms as listed below. But how many actually need checking? Are some, in fact, inherited from R?

Closure under addition $\forall a, b \in S, a + b \in S$;

Associativity of addition $\forall a, b, c \in S, (a + b) + c = a + (b + c)$;

Additive identity $\exists 0 \in S$ such that $\forall a \in S, 0 + a = a + 0 = 0$;

Additive inverses $\forall a \in S, \exists (-a) \in S$ such that
$\qquad a + (-a) = (-a) + a = 0$;

Commutativity of addition $\forall a, b \in S, a + b = b + a$;

Closure under multiplication $\forall a, b \in S, a \cdot b \in S$;

Associativity of multiplication $\forall a, b, c \in S, (a \cdot b) \cdot c = a \cdot (b \cdot c)$;

Multiplicative identity $\exists 1 \in S$ such that $\forall a \in S, 1 \cdot a = a \cdot 1 = a$;

Left distributivity $\forall a, b, c \in S, a \cdot (b + c) = a \cdot b + a \cdot c$;

Right distributivity $\forall a, b, c \in S, (a + b) \cdot c = a \cdot c + b \cdot c$.

The identity, inverses and closure axioms cannot be inherited: removing elements from R might remove an identity or some inverses or some sums or products. But the remaining axioms are inherited because each is defined using a criterion with a single universal quantifier (see Section 3.5). This means that five checks are required.

Closure under addition $\forall a, b \in S, a + b \in S$;

Additive identity $\exists 0 \in S$ such that $\forall a \in S, 0 + a = a + 0 = 0$;

Additive inverses $\forall a \in S, \exists (-a) \in S$ such that
$\qquad a + (-a) = (-a) + a = 0$;

Closure under multiplication $\forall a, b \in S, a \cdot b \in S$;

Multiplicative identity $\exists 1 \in S$ such that $\forall a \in S, 1 \cdot a = a \cdot 1 = a$.

Personally, I would be satisfied with that, as a five-item checklist is not too arduous. But mathematicians like to minimize work, so you will likely see a theorem like that below.

Theorem: Suppose that R is a ring and that $S \subseteq R$. Then S is a subring of R if $1_R \in S$ and $\forall a, b \in S$, $ab \in S$ and $a - b \in S$.

Can you work out why this is enough to guarantee that S satisfies all five axioms on the checklist (and therefore all of the ring axioms)? When you have considered that, try applying the theorem to establish that one subring of $M_{2 \times 2}(\mathbb{R})$ comprises all matrices of the form

$$\begin{pmatrix} x & 0 \\ 0 & y \end{pmatrix} \text{ where } x, y \in \mathbb{R}.$$

Now, if you have studied group theory or read Chapter 7 then you will know that some groups have 'special' subgroups known as *normal subgroups*, the *cosets* of which form a *quotient group*. (If you have not read Chapter 7, I recommend doing so now—the following text will be easier if you know how I introduced these constructs.) Is there an analogy for rings? Do some rings have 'special' subrings, the 'cosets' of which form a 'quotient ring'? I think the answer is not obvious. On the one hand, every ring is an additive group—indeed, an abelian (commutative) one. On the other hand, every ring also has multiplication. Might that mean that there are no quotient rings? Or that only some rings have quotient rings?

The answer is that quotient rings do exist. They occur when an additive subgroup is an *ideal*, where ideals are commonly defined in one of the following ways.[3]

Definition: Let $(R, +, \cdot)$ be a ring and $(S, +)$ be a subgroup of $(R, +)$. Then S is an **ideal** of R if and only if $\forall s \in S$ and $\forall r \in R$, $rs \in S$ and $sr \in S$.

Definition: Let $(R, +, \cdot)$ be a ring and $(S, +)$ be a subgroup of $(R, +)$. Then S is an **ideal** of R if and only if $\forall r \in R$, $rS \subseteq S$ and $Sr \subseteq S$.

---

[3] Other possibilities exist. *Left ideal* and *right ideal* can be defined separately, for instance.

Probably your lecturer will give a definition and go on to prove that ideals give rise to quotient rings. If you like formal theory building, you might be perfectly happy with that. If you are like me, though, you might find it unsatisfying. The above definitions are equivalent—can you see why? And they are somewhat like definitions of normal subgroup, as below.

Definition: Let H be a subgroup of G. Then H is a **normal subgroup** if and only if $\forall g \in G, gH = Hg$.

Definition: Let H be a subgroup of G. Then H is a **normal subgroup** if and only if $\forall a \in G$ and $\forall h \in H, a^{-1}ha \in H$.

But the two sets of definitions are not exactly alike. The first normal subgroup definition, for instance, requires that for every group element g, the left and right cosets gH and Hg are equal. It does not require that either coset is a subset of H, so it does not exactly match the ideal criterion that $rS \subseteq S$ and $Sr \subseteq S$. Any analogy between normal subgroups and ideals is not so direct as to be obvious.

Also not obvious is that in the normal subgroup definitions, the operation will be whatever is the single group operation, but in the ideal definition, *it must be multiplication*. This is an important distinction, and to understand its impact we need to distinguish the two operations, understand why every additive subgroup of a ring is normal, and establish conditions under which the cosets of an additive normal subgroup function both additively and multiplicatively as elements in a quotient ring. We will work through that in this section and the next, using two examples to highlight what happens when everything works and what goes wrong when it doesn't.

First, here is the definition of normal subgroup in additive form.

Definition: Let $(H, +)$ be a subgroup of $(G, +)$. Then H is a **normal subgroup** if and only if $\forall g \in G, g + H = H + g$.

The equation $g + H = H + g$ specifies that the left and right cosets of $(H, +)$ in $(G, +)$ are equal, where cosets are defined as below

Definition: Suppose that $(H, +)$ is a subgroup of $(G,)$. Then the **left coset** of H containing g is $g + H = \{g + h | h \in H\}$.

Definition: Suppose that $(H, +)$ is a subgroup of $(G, +)$. Then the **right coset** of H containing g is $H + g = \{h + g | h \in H\}$.

In an abelian group with commutative addition, the equation $g + H = H + g$ holds automatically because $g + h = h + g$ for every $h \in H$ and $g \in G$ (see Section 7.6). And every ring is an abelian group under addition. So, if $(S, +)$ is a subgroup of $(R, +)$, then $(S, +)$ must be a normal subgroup of $(R, +)$. For instance, $(3\mathbb{Z}, +)$ is a normal subgroup of $(\mathbb{Z}, +)$. And $(\mathbb{Z}, +)$ is a normal subgroup of $(\mathbb{Q}, +)$.

Consequently, the cosets in each case form an additive quotient group (see Section 7.5). As noted in Section 7.2, in $G = (\mathbb{Z}, +)$, the cosets of $H = (3\mathbb{Z}, +)$ are

$$0 + 3\mathbb{Z} = \{0 + z | z \in 3\mathbb{Z}\} = \{\ldots, -6, -3, 0, 3, 6, 9, \ldots\};$$
$$1 + 3\mathbb{Z} = \{1 + z | z \in 3\mathbb{Z}\} = \{\ldots, -5, -2, 1, 4, 7, 10, \ldots\};$$
$$2 + 3\mathbb{Z} = \{2 + z | z \in 3\mathbb{Z}\} = \{\ldots, -4, -1, 2, 5, 8, 11, \ldots\}.$$

Addition of these cosets is well defined: as in Section 7.5, it is meaningful to write $(a + 3\mathbb{Z}) + (b + 3\mathbb{Z}) = (a + b) + 3\mathbb{Z}$. For instance, adding any element of the coset $1 + 3\mathbb{Z}$ (a number with remainder 1 on division by 3) to any element of the coset $2 + 3\mathbb{Z}$ (a number with remainder 2 on division by 3) gives an element of the coset $3\mathbb{Z}$ (a number with remainder 0 on division by 3). And the cosets of $3\mathbb{Z}$ in $\mathbb{Z}$ behave 'as they should' under addition,[4] forming a quotient group $\mathbb{Z}/3\mathbb{Z} \cong \mathbb{Z}_3$.

$+_3$	$3\mathbb{Z}$	$1 + 3\mathbb{Z}$	$2 + 3\mathbb{Z}$
$3\mathbb{Z}$	$3\mathbb{Z}$	$1 + 3\mathbb{Z}$	$2 + 3\mathbb{Z}$
$1 + 3\mathbb{Z}$	$1 + 3\mathbb{Z}$	$2 + 3\mathbb{Z}$	$3\mathbb{Z}$
$2 + 3\mathbb{Z}$	$2 + 3\mathbb{Z}$	$3\mathbb{Z}$	$1 + 3\mathbb{Z}$

$+_3$	0	1	2
0	0	1	2
1	1	2	0
2	2	0	1

[4] See Section 3.3 for a more formal argument for cosets of $(12\mathbb{Z}, +)$ in $(\mathbb{Z}, +)$.

Do these cosets also behave 'as they should' under multiplication, thus forming a quotient *ring*? They do. The tables below represent the same additive cosets under the multiplicative operation $\times_3$.

$\times_3$	$3\mathbb{Z}$	$1+3\mathbb{Z}$	$2+3\mathbb{Z}$
$3\mathbb{Z}$	$3\mathbb{Z}$	$3\mathbb{Z}$	$3\mathbb{Z}$
$1+3\mathbb{Z}$	$3\mathbb{Z}$	$1+3\mathbb{Z}$	$2+3\mathbb{Z}$
$2+3\mathbb{Z}$	$3\mathbb{Z}$	$2+3\mathbb{Z}$	$1+3\mathbb{Z}$

$\times_3$	0	1	2
0	0	0	0
1	0	1	2
2	0	2	1

The structure is different, but multiplication is well defined (compare with Section 3.3). For instance, any element of the coset $1+3\mathbb{Z}$ (a number with remainder 1 on division by 3) multiplied by any element of the coset $2+3\mathbb{Z}$ (a number with remainder 2 on division by 3) gives an element of the coset $2+3\mathbb{Z}$ (a number with remainder 2 on division by 3). Thus it is meaningful to write $(1+3\mathbb{Z})(2+3\mathbb{Z}) = 2+3\mathbb{Z}$ and, in general,

$$(a+3\mathbb{Z})(b+3\mathbb{Z}) = (ab)+3\mathbb{Z}.$$

In ring theoretic terms, this is what we would expect, because $3\mathbb{Z}$ satisfies the definition of an ideal in $\mathbb{Z}$. One formulation of this definition is repeated below and applied to $3\mathbb{Z}$ as a subring of $\mathbb{Z}$.

Definition: Let $(R, +, \cdot)$ be a ring and $(S, +)$ be a subgroup of $(R, +)$. Then S is an **ideal** of R if and only if $\forall s \in S$ and $\forall r \in R$, $rs \in S$ and $sr \in S$.

Application: $3\mathbb{Z}$ is an additive subgroup of $\mathbb{Z}$; it is an ideal of $\mathbb{Z}$ because $\forall s \in 3\mathbb{Z}$ and $\forall r \in \mathbb{Z}$, $rs \in 3\mathbb{Z}$ and $sr \in 3\mathbb{Z}$.

In words, the definition means that multiplying any element of an ideal S by any element of the ring R gives a result in S; informally, S 'drags everything into itself' under multiplication. For $3\mathbb{Z}$ as a subring of $\mathbb{Z}$, multiplying any multiple of 3 by any integer gives a multiple of 3. The above tables might convince you that $\mathbb{Z}/3\mathbb{Z}$ is a meaningful quotient ring isomorphic to $(\mathbb{Z}_3, +_3, \times_3)$. But what happens if a subring is not an ideal?

9.7 Ideals, quotient rings and ring homomorphisms

The previous section established that $3\mathbb{Z}$ is an ideal of $\mathbb{Z}$, and that its cosets form not only a quotient group under addition but also a quotient ring under addition and multiplication. So that is an example where everything 'works'. But what of the second example? Is $\mathbb{Z}$ an ideal of $\mathbb{Q}$? No, because it is not true that $\forall z \in \mathbb{Z}$ and $\forall q \in \mathbb{Q}$, $zq \in \mathbb{Z}$ and $qz \in \mathbb{Z}$. For instance, multiplying $2 \in \mathbb{Z}$ by $\frac{5}{7} \in \mathbb{Q}$ gives $\frac{10}{7} \notin \mathbb{Z}$. In theory-building terms, we could shrug and say 'Okay, no quotient ring, then'. But I do not find that satisfying. I want to know *why* $\mathbb{Z}$ failing to be an ideal of $\mathbb{Q}$ means that there is no quotient ring $\mathbb{Q}/\mathbb{Z}$. To find out, we will work through some reasoning.

Because $(\mathbb{Z}, +)$ is a normal subgroup of $(\mathbb{Q}, +)$, there is a quotient *group* $\mathbb{Q}/\mathbb{Z}$. Its elements are the additive cosets of $\mathbb{Z}$ in $\mathbb{Q}$, which take the form $q + \mathbb{Z}$ where $q \in \mathbb{Q}$. These cosets are infinite in number, so we cannot list them all. But here are a few.

$$0 + \mathbb{Z} = \{0 + z | z \in \mathbb{Z}\} = \{\ldots, -2, -1, 0, 1, 2, \ldots\};$$
$$\tfrac{1}{4} + \mathbb{Z} = \{\tfrac{1}{4} + z | z \in \mathbb{Z}\} = \{\ldots, -\tfrac{7}{4}, -\tfrac{3}{4}, \tfrac{1}{4}, \tfrac{5}{4}, \tfrac{9}{4}, \ldots\};$$
$$\tfrac{1}{2} + \mathbb{Z} = \{\tfrac{1}{2} + z | z \in \mathbb{Z}\} = \{\ldots, -\tfrac{3}{2}, -\tfrac{1}{2}, \tfrac{1}{2}, \tfrac{3}{2}, \tfrac{5}{2}, \ldots\}.$$

Because $(\mathbb{Z}, +)$ is a normal subgroup of $(\mathbb{Q}, +)$, coset addition is well defined. For instance, adding an element of $\frac{1}{4} + \mathbb{Z}$ (a number with 'remainder' $\frac{1}{4}$ on division by 1) to any element of $\frac{1}{2} + \mathbb{Z}$ (a number with 'remainder' $\frac{1}{2}$ on division by 1) gives an element of $\frac{3}{4} + \mathbb{Z}$ (a number with 'remainder' $\frac{3}{4}$ on division by 1). In general,

$$(q_1 + \mathbb{Z}) + (q_2 + \mathbb{Z}) = (q_1 + q_2) + \mathbb{Z}.$$

Obviously infinite coset addition tables are not possible, but below are some partial ones. Perhaps pick a few more cosets and add those too.

+	$\mathbb{Z}$	$\frac{1}{4}+\mathbb{Z}$	$\frac{1}{2}+\mathbb{Z}$	$\cdots$
$\mathbb{Z}$	$\mathbb{Z}$	$\frac{1}{4}+\mathbb{Z}$	$\frac{1}{2}+\mathbb{Z}$	
$\frac{1}{4}+\mathbb{Z}$	$\frac{1}{4}+\mathbb{Z}$	$\frac{1}{2}+\mathbb{Z}$	$\frac{3}{4}+\mathbb{Z}$	
$\frac{1}{2}+\mathbb{Z}$	$\frac{1}{2}+\mathbb{Z}$	$\frac{3}{4}+\mathbb{Z}$	$\mathbb{Z}$	
$\vdots$				

+	0	$\frac{1}{4}$	$\frac{1}{2}$	$\cdots$
0	0	$\frac{1}{4}$	$\frac{1}{2}$	
$\frac{1}{4}$	$\frac{1}{4}$	$\frac{1}{2}$	$\frac{3}{4}$	
$\frac{1}{2}$	$\frac{1}{2}$	$\frac{3}{4}$	0	
$\vdots$				

Now, because $\mathbb{Z}$ is not an ideal of $\mathbb{Q}$, we are expecting that coset multiplication will not be well defined—that $(q_1 + \mathbb{Z})(q_2 + \mathbb{Z})$ will not be meaningfully equal to $(q_1 q_2) + \mathbb{Z}$. But what exactly goes wrong? Imagine multiplying $\frac{1}{4} + \mathbb{Z}$ by $\frac{1}{2} + \mathbb{Z}$. What should the answer be? Using the obvious coset representatives $\frac{1}{4} \in \frac{1}{4} + \mathbb{Z}$ and $\frac{1}{2} \in \frac{1}{2} + \mathbb{Z}$ gives

$$\frac{1}{4} \cdot \frac{1}{2} = \frac{1}{8} \in \frac{1}{8} + \mathbb{Z} \text{ so we would want } (\frac{1}{4} + \mathbb{Z}) \cdot (\frac{1}{2} + \mathbb{Z}) = \frac{1}{8} + \mathbb{Z}.$$

But taking the alternative representatives $\frac{5}{4} \in \frac{1}{4} + \mathbb{Z}$ and $\frac{7}{2} \in \frac{1}{2} + \mathbb{Z}$ gives

$$\frac{5}{4} \cdot \frac{7}{2} = \frac{35}{8} \in \frac{3}{8} + \mathbb{Z} \text{ so we would want } (\frac{1}{4} + \mathbb{Z}) \cdot (\frac{1}{2} + \mathbb{Z}) = \frac{3}{8} + \mathbb{Z}.$$

Because $\frac{1}{8} + \mathbb{Z} \neq \frac{3}{8} + \mathbb{Z}$, coset multiplication is not well defined (you can check that it fails similarly for many other cosets and representatives). Thus coset multiplication is not meaningful and $\mathbb{Q}/\mathbb{Z}$ is not a quotient ring.

That is what we expected, but it does not really explain why a quotient ring requires an ideal. To understand that, we will clarify what goes wrong when multiplying $\frac{1}{4} + \mathbb{Z}$ by $\frac{1}{2} + \mathbb{Z}$, then generalize. Consider again the coset representatives $\frac{5}{4} \in \frac{1}{4} + \mathbb{Z}$ and $\frac{7}{2} \in \frac{1}{2} + \mathbb{Z}$, and recall that all elements in a coset are 'separated from one another' by elements of the subgroup (see Section 7.6). Here, $\frac{5}{4} = \frac{1}{4} + 1$ where $1 \in \mathbb{Z}$ and $\frac{7}{2} = \frac{1}{2} + 3$ where $3 \in \mathbb{Z}$, so we can rewrite the product as

$$\frac{5}{4} \cdot \frac{7}{2} = (\frac{1}{4} + 1)(\frac{1}{2} + 3) = \frac{1}{8} + \frac{3}{4} + \frac{1}{2} + 3.$$

If the result were simply $\frac{1}{8} + 3$, it would be an element of $\frac{1}{8} + \mathbb{Z}$. But multiplying also manufactures 'extra bits', in this case $\frac{3}{4}$ and $\frac{1}{2}$.

In general, representing $\frac{1}{4} + \mathbb{Z}$ by $\frac{1}{4} + z_1$ where $z_1 \in \mathbb{Z}$ and $\frac{1}{2} + \mathbb{Z}$ by $\frac{1}{2} + z_2$ where $z_2 \in \mathbb{Z}$ gives

$$(\tfrac{1}{4} + z_1)(\tfrac{1}{2} + z_2) = \tfrac{1}{8} + \tfrac{1}{4}z_2 + \tfrac{1}{2}z_1 + 1.$$

The result is an element of $\frac{1}{8} + \mathbb{Z}$ only if the extra bits $\frac{1}{4}z_2$ and $\frac{1}{2}z_1$ are integers. Are they? Not usually. And for coset multiplication to be well defined, they would need to be integers in every case. Do you see the link to the definition of ideal?

Definition: Let $(R, +, \cdot)$ be a ring and $(S, +)$ be a subgroup of $(R, +)$. Then S is an **ideal** of R if and only if $\forall s \in S$ and $\forall r \in R, rs \in S$ and $sr \in S$.

Definition: Let $(R, +, \cdot)$ be a ring and $(S, +)$ be a subgroup of $(R, +)$. Then S is an **ideal** of R if and only if $\forall r \in R$, $rS \subseteq S$ and $Sr \subseteq S$.

To nail the reasoning, consider the fully general case. Suppose that R is a ring and that $(S, +)$ is a subgroup of $(R, +)$, so that cosets of S in R take the form $r_1 + S, r_2 + S$ and so on; these cosets might be finite or infinite, depending on the ring.

$$r_1 + S = \{r_1,\ r_1 + s_1,\ r_1 + s_2, \ldots\};$$
$$r_2 + S = \{r_2,\ r_2 + s_1,\ r_2 + s_2, \ldots\};$$
$$r_3 + S = \{r_3,\ r_3 + s_1,\ r_3 + s_2, \ldots\};$$
$$\vdots$$

Adding the cosets $r_1 + S$ and $r_2 + S$ is unproblematic. Taking arbitrary representatives $r_1 + s_m \in r_1 + S$ and $r_2 + s_n \in r_2 + S$ gives

$$\begin{aligned}
(r_1 + s_m) + (r_2 + s_n) &= r_1 + (s_m + r_2) + s_n && \text{because addition is associative} \\
&= r_1 + r_2 + s_m + s_n && \text{because addition is commutative} \\
&= (r_1 + r_2) + (s_m + s_n) \\
&\in (r_1 + r_2) + S && \text{because } s_m + s_n \in S.
\end{aligned}$$

Thus it is meaningful to write $(r_1 + S) + (r_2 + S) = (r_1 + r_2) + S$.

But *multiplying* the cosets $r_1 + S$ and $r_2 + S$ is problematic. Taking again the representatives $r_1 + s_m$ and $r_2 + s_n$ and using distributivity gives

$$(r_1 + s_m)(r_2 + s_n) = r_1 r_2 + r_1 s_n + s_m r_2 + s_m s_n.$$

Because S is a subring, it is closed under multiplication; hence $s_m s_n \in S$. But it is meaningful to write $(r_1 + S)(r_2 + S) = (r_1 r_2) + S$ only if $r_1 s_n$ and $s_m r_2$ are definitely also in S. In other words, S must be an ideal.

We are nearly done now, but not quite. The above reasoning establishes that for R/S to be a quotient ring, it is necessary that S be an ideal. Is it also sufficient? Does every ideal give rise to a quotient ring? The answer, happily, is yes, but we have not yet established that. We know that if S is an ideal then coset addition and multiplication are well defined, but we have not proved that these operations obey the ring axioms. Proving that they do is not that difficult but also not that interesting, partly because there are so many axioms. Your course might offer a full proof; here I will illustrate with two.

First, consider closure under addition. Here is the axiom, in its general form and translated into a claim about cosets in R/S.

Closure under addition $\forall a, b \in R, a + b \in R$;
Closure under addition $\forall a + S, b + S \in R/S$,
    $(a + S) + (b + S) \in R/S$.

Cosets obey this axiom because, as we have established, $(a + S) + (b + S) = (a + b) + S$, which is a coset of S in R and is thus an element of R/S.

Second, consider left distributivity.

Left distributivity $\forall a, b, c \in R, a \cdot (b + c) = a \cdot b + a \cdot c$;
Left distributivity $\forall a + S, b + S, c + S \in R/S$,
    $(a + S) \cdot ((b + S) + (c + S)) = (a + S) \cdot (b + S) + (a + S) \cdot (c + S)$.

Cosets obey this axiom due to the argument below. Where does this use properties of coset addition and multiplication, and where does it use properties of ring addition and multiplication?

$$\begin{aligned}(a+S)\cdot((b+S)+(c+S)) &= (a+S)\cdot((b+c)+S)\\ &= (a\cdot(b+c))+S\\ &= (a\cdot b+a\cdot c)+S\\ &= (a\cdot b+S)+(a\cdot c+S)\\ &= (a+S)\cdot(b+S)+(a+S)\cdot(c+S).\end{aligned}$$

As I say, such proofs are not that interesting. But working through the list of axioms proves that if S is an ideal of R then R/S is a ring, and provides useful exercise in taking care over algebraic validity (thus linking right back to Chapter 1).

To relate the theory to examples, we can consider more ideals and non-ideals. In some rings, every additive subgroup is an ideal. For instance, in $\mathbb{Z}_6$ all additive subgroups are ideals—why? Does this generalize to $\mathbb{Z}_n$? Other rings have more 'room' for substructures that are not ideals. As above, $\mathbb{Z}$ is an additive subgroup—in fact a subring—of $\mathbb{Q}$, but not an ideal. Similarly, $\mathbb{Q}$ is a subring but not an ideal of $\mathbb{R}$: for instance, $1 \in \mathbb{Q}$ and $\sqrt{2} \in \mathbb{R}$ but $1\sqrt{2} \notin \mathbb{Q}$. The set

$$S = \left\{\begin{pmatrix} x & 0 \\ 0 & y \end{pmatrix} | x,y \in \mathbb{R}\right\}$$

is a subring but not an ideal of $M_{2\times2}(\mathbb{R})$, because multiplying one of its elements by a general element of $M_{2\times2}(\mathbb{R})$ gives

$$\begin{pmatrix} x & 0 \\ 0 & y \end{pmatrix}\begin{pmatrix} a & b \\ c & d \end{pmatrix} = \begin{pmatrix} xa & xb \\ yc & yd \end{pmatrix},$$

which need not be in S. Can you find an ideal of $M_{2\times2}(\mathbb{R})$, or convince yourself that this is impossible? How about ideals in polynomial rings? For the non-ideals, where does the above argument about distributivity break down?

To conclude, some brief comments about where your course will likely go with these ideas, which is to *ring homomorphims* and *ring isomorphisms*. The group versions of these constructs were discussed in Chapter 8. Group homomorphisms are maps between groups that respect their operations; group isomorphisms are bijective homomorphisms.

Definition: $\phi : G_1 \to G_2$ is a **group homomorphism** if and only if $\forall a, b \in G_1, \phi(ab) = \phi(a)\phi(b) \in G_2$.

Definition: $\phi : G_1 \to G_2$ is a **group isomorphism** if and only if ϕ is bijective and $\forall a, b \in G_1, \phi(ab) = \phi(a)\phi(b) \in G_2$.

Ring homomorphisms and isomorphisms are analogous, with the obvious adjustment that they respect both ring operations.

Definition: $\phi : R_1 \to R_2$ is a **ring homomorphism** if and only if $\forall a, b \in R_1, \ \phi(a+b) = \phi(a) + \phi(b) \in R_2$ and $\phi(a \cdot b) = \phi(a) \cdot \phi(b) \in R_2$.

Definition: $\phi : R_1 \to R_2$ is a **ring isomorphism** if and only if ϕ is bijective and $\forall a, b \in R_1, \phi(a+b) = \phi(a) + \phi(b) \in R_2$ and $\phi(a \cdot b) = \phi(a) \cdot \phi(b) \in R_2$.

As with the group versions, the criteria $\phi(a+b) = \phi(a) + \phi(b)$ and $\phi(a \cdot b) = \phi(a) \cdot \phi(b)$ can be understood in terms of order of operations: adding then mapping gives the same result as mapping then adding, and multiplying then mapping gives the same result as mapping then multiplying.

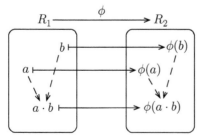

Moreover, as with the group versions, ring homomorphisms link properties of elements via theorems like those below. Can you adapt ideas from Chapter 8 to construct proofs?

Theorem: Suppose $\phi : R_1 \to R_2$ is a ring homomorphism.
Then $\phi(0_{R_1}) = 0_{R_2}$.

Theorem: Suppose that $\phi : R_1 \to R_2$ is a ring homomorphism.
Then $\forall a \in R_1, \phi(-a) = -\phi(a)$.

Ring homomorphisms also have *kernels*, where the kernel of $\phi : R_1 \to R_2$ is the subset of R_1 containing all elements that map to the additive identity. The definition and theorem below are analogous to those for groups. Can you adapt ideas from Section 8.7 to prove the theorem?

Definition: Suppose that $\phi : R_1 \to R_2$ is a ring homomorphism. Then the **kernel** of ϕ is $\ker \phi = \{k \in R_1 | \phi(k) = 0_{R_2}\}$.

Theorem: Suppose that $\phi : R_1 \to R_2$ is a ring homomorphism and that $\ker \phi = \{0_{R_1}\}$. Then ϕ is injective.

Finally, with the definition of kernel in place, we can state two 'bigger' theorems. Imagery from Section 8.7 should help in thinking about what they mean.

Theorem: Suppose that $\phi : R_1 \to R_2$ is a ring homomorphism.
Then $\ker \phi$ is an ideal of R_1.

Theorem: Suppose that $\phi : R_1 \to R_2$ is a ring homomorphism.
Then $R_1 / \ker \phi \cong \mathrm{im}\phi$.

If your course involves a lot of ring theory, you will likely study all of these theorems in depth. You might also prove that every ideal is the kernel of a ring homomorphism, which provides a theory-based way to establish that every ideal gives rise to a quotient ring—can you see see why? If your institution covers group theory and ring theory in different courses, analogies across the two might not be apparent, so I recommend thinking about them now. Either way, these theorems conclude both our introduction to rings and this book's main content.

CONCLUSION

This concluding chapter begins with a brief review of the mathematical concepts covered in Part 2 of the book. It then revisits ideas from Part 1, suggesting useful things to bear in mind when studying Abstract Algebra.

The mathematical content in Part 2 of this book began with binary operations. Chapter 5 treated these as a separate topic because research indicates that students might not give them the attention necessary to understand their properties. It contrasted associativity with commutativity, noting that these need not necessarily co-occur, then discussed binary operations in modular arithmetic and on functions, matrices, transformations, symmetries and permutations. It established that function composition is always associative—a result that applies across many structures—and discussed technical issues around closure.

Chapter 6 discussed groups and subgroups, noting that the definition of a group can be expressed in various notations. It introduced subgroups as subsets of groups that are groups in their own right. It then explored cyclic groups, discussing generators, commutativity, and the integers as an infinite cyclic group. It explained groups and subgroups that exist in familiar structures like the rational, real and complex numbers, and sets of matrices. It discussed the dihedral group D_3—the group of symmetries of an equilateral triangle—along with other dihedral and symmetry groups and ways in which they differ from cyclic groups. Finally, it discussed groups of permutations, ways to identify and define subgroups, and structures of small groups.

Chapter 7 began by noting that some groups split naturally into a subgroup and cosets that form elements in a quotient group. It explored cases in which this happens and cases in which it does not, observing that cosets always partition the group but that a quotient group arises if and only if the left and right cosets are the same. It discussed definitions of normal subgroup, and observed that some subgroups are guaranteed to be normal due to properties involving commutativity. It observed that within a coset, every element is 'offset' from the subgroup by the same 'remainder', and that elements differ by elements of the subgroup. Finally, it discussed Lagrange's Theorem, which restricts possible subgroup orders in finite groups.

Chapter 8 discussed isomorphic groups as those that are structurally identical, and isomorphisms as bijective functions between isomorphic groups. It observed that the criterion $\phi(ab) = \phi(a)\phi(b)$ can be understood in terms of order of operations, and that this criterion imposes tight restrictions on which functions are isomorphisms. It related isomorphisms to commutativity and identities, later observing that the proofs do not require bijectivity so apply to homomorphisms too. It provided examples of isomorphisms and homomorphisms, and noted that homomorphisms can be understood as organizing elements into sets in which every element maps to the same element of the image. This idea was then formalized in the First Isomorphism Theorem.

Finally, Chapter 9 introduced rings as sets with two binary operations, addition and multiplication. It provided examples of rings of numbers, congruence classes, matrices and polynomials, and discussed simple theorems linking additive identities and inverses to the multiplicative ring operation. It discussed additional conditions that a ring must satisfy to be commutative, a division ring, an integral domain, or a field, and explored properties of units and zero divisors. It then explained why a quotient ring exists if and only if an additive subgroup is an ideal. It concluded by noting parallels between ring theory and group theory.

Overall, I hope that Part 2 conveyed the idea that Abstract Algebra involves recognizing deep similarities across a range of structures. I think that this makes Abstract Algebra the most obvious place to understand what the mathematician Poincaré meant in saying that mathematics is the art of giving the same name to different things. Do you agree?

Of course, this book is just a start. I have aimed to introduce key concepts in depth, providing information on ways to think about these accurately and productively. But great swathes of Abstract Algebra are not covered here, and there is considerable variety in Abstract Algebra courses. If your lecturer takes a formal approach, you might cover most of the ideas in this book in a small number of weeks. If your lecturer takes a geometric approach, the emphasis might be on symmetries and permutations, and you might only later explore general theorems and applications to other structures. Whatever happens, you will cover material not in this book.

To learn effectively, it is therefore worth revisiting ideas from Part 1. If you are new to undergraduate (or upper-level) pure mathematics, you might have little experience with axioms, definitions, theorems and proofs, so the information in Chapters 2 and 3 might have been new. If you have already studied at this level, it might still be worth reflecting on your study habits. Do you give definitions the attention they deserve, applying them to examples and noting where they are used in proving theorems? Do you understand what a theorem says before you try to read its proof or construct your own? When reading mathematics, do you aim for good self-explanations? Most students do all of these things, just not consistently—usually for the ordinary human reasons that they are tired or in a rush. But it is important to focus on them, because the implicit messages from textbooks, lectures and the passage of time tend to push students in unproductive directions.

Textbooks can be used ineffectively if students think of them only as a source of exercises. This would not be their fault: teachers sometimes treat textbooks that way. But virtually all textbooks contain good explanations in their expository sections. These become more important in undergraduate study, where you are unlikely to understand everything in lectures: a slightly different explanation might make the links you need. That makes it worth investing an hour or two in finding a book that you find helpful. I recommend going to your university's library and assembling five or six Abstract Algebra books. Find those recommended by your lecturer, then scan the shelves for any of which there are multiple copies—someone, at some point, thought these worth ordering. Then, in each book, find the beginning of the same topic—groups, say—and read the first couple of pages. This will reveal big differences in expository style: some books are terse and others are wordy, some have many examples and others have few, and so on. Some you will immediately dislike. But one or two you will find useful, perhaps in different ways. Those are the ones to borrow or buy, and to study regularly.

Lectures, too, can discourage good habits. They make it is easy to act like an automaton, copying down what the lecturer writes without really taking it in. This is not your fault—the passive nature of many lectures makes it hard, at times, to do better. But it *is* your responsibility. No matter what you think of your course, the only person who can do the learning is you. So you have to work out how to learn effectively from whatever is happening in the room. One specific problem is that lecture notes often contain only the bare-bones axioms, definitions, theorems and proofs. Much of the useful information is in what the lecturer says but does *not* write down. Lecturers explain the thinking that links one line of a proof to another, refer back to earlier ideas, and so on. Recording this information might be difficult: in some lectures, students can barely keep up. But often that is not the case. Many lecturers provide notes or lecture recordings online, which frees you up to make choices about whether to try to write everything, or to pre-print notes and annotate them, or to make minimal notes and rewatch difficult sections later. Certainly your notes do not have to look the same as everyone else's. If you take down whatever seems useful for your thinking, you are putting in self-explanations as you go.

Finally, good habits are hard to maintain when you are getting a bit panicky because your course is moving fast and you are getting behind. In those circumstances, people often try to speed up, glossing over things that they don't really understand in order to 'catch up'. But usually they need to do the opposite. Patchy understanding isn't much use in a subject like Abstract Algebra, and you will often make more progress in a given week if you first sort out your understanding of its most important ideas. And that means you do need to be studying each week. As discussed in Chapter 1, Abstract Algebra is very hierarchical. This makes it particularly unforgiving of poor study habits: students who do not study consistently can soon find that it makes no sense at all. To avoid that situation, I recommend reviewing Chapter 4.

I would then like to conclude by drawing together a thread that has run through this book. Meaningful understanding is important, and my top advice for developing meaningful understanding would be to ask, *what type of object is that?* Does that symbol denote a single element or a set? If it denotes a single element, what type of object is it? Is it a number, perhaps, or a transformation? Is it a special kind of transformation like a symmetry? Is the specific object important in its context, in the sense that it has special properties? Or is it an arbitrary member of a group or ring? If a symbol denotes a set, is that set unstructured, or is it a group or a ring? Is it treated as a subset or subgroup or subring of another set, group or ring? Is it, in fact, a coset of a subgroup within a group? If so, is the subgroup normal, so that this coset and others form a quotient group? Is it a subring, or maybe an ideal? Do we know anything else about its properties? Does it have inverses, or an identity, or commutativity under some binary operation?

Asking these questions should help you toward meaningful understanding: identifying the objects in an axiom, definition or theorem will clarify its meaning, especially if paired with explicit thought about logic, as discussed in Section 3.5. This should facilitate self-explanation, as described in Section 3.6. It should also facilitate proof construction; Section 3.7 provided some thoughts on that but, if you read it before Part 2, you now know much more about a range of concepts, so it would be worth revisiting. It is then worth thinking about why a meaningful understanding provides a strong base for constructing proofs and for doing well in Abstract Algebra. I would list three reasons. First, when trying to prove something, you will often have access to a lot of possibly useful axioms, definitions and earlier theorems; deciding which to use is much easier with clear understanding of what each one says. Second, the theorems of Abstract Algebra are all expressed in similar notation; students who do not really understand them can easily muddle them up. Third, it is difficult to remember things that are not meaningfully understood. I certainly would not want to attempt an Abstract Algebra exam based on brute-force memorization. I am sure it can be done, but I have long thought that students who get the top marks probably work on average less hard than those who get slightly lower marks, because they are doing a different and, in the long run, slightly easier task. Acquiring meaningful understanding requires a lot of up-front effort, but meaningful understanding sticks.

BIBLIOGRAPHY

Ainsworth, S., & Loizou, A. T. (2003). The effects of self-explanation when learning with text or diagrams. *Cognitive Science, 27*, 669–681.

Alcock, L. (2010). Mathematicians' perspectives on the teaching and learning of proof. In F. Hitt, D. Holton, & P. W. Thompson (Eds.), *Research in collegiate mathematics education vii* (pp. 63–92). Washington DC: MAA.

Alcock, L. (2013a). *How to study as a mathematics major.* Oxford: Oxford University Press.

Alcock, L. (2013b). *How to study for a mathematics degree.* Oxford: Oxford University Press.

Alcock, L. (2014). *How to think about analysis.* Oxford: Oxford University Press.

Alcock, L., Bailey, T., Inglis, M., & Docherty, P. (2014). The ability to reject invalid logical inferences predicts proof comprehension and mathematics performance. In *Proceedings of the 17th conference on research in undergraduate mathematics education.* Denver, CO.

Alcock, L., Brown, G., & Dunning, T. C. (2015). Independent study workbooks for proofs in group theory. *International Journal of Research in Undergraduate Mathematics Education, 1*, 3–26.

Alcock, L., Hodds, M., Roy, S., & Inglis, M. (2015). Investigating and improving undergraduate proof comprehension. *Notices of the American Mathematical Society, 62*, 742–752.

Alcock, L., & Inglis, M. (2008). Doctoral students' use of examples in evaluating and proving conjectures. *Educational Studies in Mathematics, 69*, 111–129.

Alcock, L., & Simpson, A. (2002). Definitions: dealing with categories mathematically. *For the Learning of Mathematics, 22*(2), 28–34.

Alcock, L., & Weber, K. (2010a). Referential and syntactic approaches to proving: Case studies from a transition-to-proof course. In F. Hitt, D. Holton,

& P. W. Thompson (Eds.), *Research in collegiate mathematics education vii* (p. 93–114). Washington, DC: MAA.

Alcock, L., & Weber, K. (2010b). Undergraduates' example use in proof construction: Purposes and effectiveness. *Investigations in Mathematics Learning*, *3*, 1–22.

Allenby, R. B. J. T. (1991). *Rings, fields and groups: An introduction to abstract algebra*. Oxford: Butterworth Heinemann.

Asghari, A. H., & Khosroshahi, L. G. (2017). Making associativity operational. *International Journal of Science and Mathematics Education*, *15*, 1559–1577.

Ash, R. B. (2000). *Basic abstract algebra for graduate students and advanced undergraduates*. Mineola, NY: Dover Publications, Inc.

Asiala, M., Brown, A., Kleiman, J., & Mathews, D. (1998). The development of students' understanding of permutations and symmetries. *International Journal of Computers for Mathematical Learning*, *3*, 13–43.

Asiala, M., Dubinsky, E., Matthews, D. W., Morics, S., & Oktac, A. (1997). Development of students' understanding of cosets, normality, and quotient groups. *Journal of Mathematical Behavior*, *16*, 241–309.

Attridge, N., Doritou, M., & Inglis, M. (2015). The development of reasoning skills during compulsory 16 to 18 mathematics education. *Research in Mathematics Education*, *17*, 20–37.

Attridge, N., & Inglis, M. (2013). Advanced mathematical study and the development of conditional reasoning skills. *PLoS ONE*, *8*, e69399.

Bansilal, S., Brijlall, D., & Trigueros, M. (2017). An APOS study on pre-service teachers' understanding of injections and surjections. *Journal of Mathematical Behavior*, *48*, 22–37.

Bass, H. (2017). Designing opportunities to learn mathematics theory-building practices. *Educational Studies in Mathematics*, *95*, 229–244.

Benjamin, A. S., Bjork, R. A., & Schwartz, B. L. (1998). The mismeasure of memory: When retrieval fluency is misleading as a metamnemonic index. *Journal of Experimental Psychology: General*, *127*(1), 55.

Bielaczyc, K., Pirolli, P. L., & Brown, A. L. (1995). Training in self-explanation and self-regulation strategies: Investigating the effects of knowledge acquisition activities on problem solving. *Cognition and Instruction*, *13*, 221–252.

Bjork, R. A., Dunlosky, J., & Kornell, N. (2013). Self-regulated learning: Beliefs, techniques, and illusions. *Annual Review of Psychology, 64*, 417–444.

Brown, A., DeVries, D. J., Dubinsky, E., & Thomas, K. (1997). Learning binary operations, groups and subgroups. *Journal of Mathematical Behavior, 16*, 187–239.

Brown, S. (2018). Difficult dialogs about degenerate cases: A proof script study. *Journal of Mathematical Behavior, 52*, 61–76.

Buchbinder, O., & Zaslavsky, O. (2011). Is this a coincidence? The role of examples in fostering a need for proof. *ZDM: The International Journal on Mathematics Education, 43*, 269–281.

Burn, B. (1996). What are the fundamental concepts of group theory? *Educational Studies in Mathematics, 31*, 371–377.

Burn, R. P. (1998). Participating in the learning of group theory. *PRIMUS, 8*, 304–316.

Cameron, P. J. (2008). *Introduction to algebra*. Oxford: Oxford University Press.

Cepeda, N. J., Pashler, H., Vul, E., Wixted, J. T., & Rohrer, D. (2006). Distributed practice in verbal recall tasks: A review and quantitative synthesis. *Psychological Bulletin, 132*(3), 354.

Chi, M. T. H., Leeuw, N. D., Chiu, M.-H., & LaVancher, C. (1994). Eliciting self-explanations improves understanding. *Cognitive Science, 18*, 439–477.

Clark, A. (1971). *Elements of abstract algebra*. New York: Dover Publications, Inc.

Conradie, J., & Frith, J. (2000). Comprehension tests in mathematics. *Educational Studies in Mathematics, 42*, 225–235.

Cook, J. P. (2014). The emergence of algebraic structure: Students come to understand units and zero-divisors. *International Journal of Mathematical Education in Science and Technology, 45*, 349–359.

Cook, J. P. (2018). An investigation of an undergraduate student's reasoning with zero-divisors and the zero-product property. *Journal of Mathematical Behavior, 49*, 95–115.

Cook, J. P., & Fukawa-Connelly, T. (2015). The pedagogical examples of groups and rings that algebraists think are most important in an introductory course. *Canadian Journal of Science, Mathematics and Technology Education, 15*, 171–185.

Cowen, C. (1991). Teaching and testing mathematics reading. *American Mathematical Monthly*, *98*, 50–53.

Crawford, K., Gordon, S., Nicholas, J., & Prosser, M. (1998). University mathematics students' conceptions of mathematics. *Studies in Higher Education*, *23*, 87–94.

Credé, M., & Kuncel, N. R. (2008). Study habits, skills, and attitudes. *Perspectives on Psychological Science*, *3*, 425–453.

Dawkins, P. C. (2017). On the importance of set-based meanings for categories and connectives in mathematical logic. *International Journal of Research in Undergraduate Mathematics Education*, *3*, 496–522.

Dawkins, P. C., & Cook, J. P. (2017). Guiding reinvention of conventional tools of mathematical logic: Students' reasoning about mathematical disjunctions. *Educational Studies in Mathematics*, *94*, 241–256.

Dawkins, P. C., & Karunakaran, S. S. (2016). Why research on proof-oriented mathematical behavior should attend to the role of particular mathematical content. *The Journal of Mathematical Behavior*, *44*, 65–75.

de Villiers, M. (1990). The role and function of proof in mathematics. *Pythagoras*, *24*, 17–24.

Diakidoy, I.-A. N., Mouskounti, T., Fella, A., & Ioannides, C. (2016). Comprehension processes and outcomes with refutation and expository texts and their contribution to learning. *Learning and Instruction*, *41*, 60–69.

Dubinsky, E., Dautermann, J., Leron, U., & Zazkis, R. (1994). On learning fundamental concepts of group theory. *Educational Studies in Mathematics*, *27*, 267–305.

Dubinsky, E., Dautermann, J., Leron, U., & Zazkis, R. (1997). A reaction to Burn's "What are the fundamental concepts of group theory?". *Educational Studies in Mathematics*, *34*, 249–253.

Dubinsky, E., & Yiparaki, O. (2000). On student understanding of AE and EA quantification. In E. Dubinsky, A. H. Schofield, & J. Kaput (Eds.), *Research in collegiate mathematics IV* (pp. 239–289). Providence: RI: American Mathematical Society.

Dummit, D. S., & Foote, R. M. (2004). *Abstract algebra*. New Delhi: John Wiley & Sons.

Durand-Guerrier, V. (2003). Which notion of implication is the right one? From logical considerations to a didactic perspective. *Educational Studies in Mathematics*, *53*, 5–34.

Durand-Guerrier, V. (2008). Truth versus validity in mathematical proof. *ZDM: The International Journal on Mathematics Education*, *40*, 373–384.

Durkin, K. (2011). The self-explanation effect when learning mathematics: A meta-analysis. Evanston IL: Society for Research on Educational Effectiveness.

Edwards, B. S., & Ward, M. B. (2004). Surprises from mathematics education research: Student (mis)use of mathematical definitions. *American Mathematical Monthly*, *111*, 411–424.

Epp, S. (2003). The role of logic in teaching proof. *American Mathematical Monthly*, *110*, 886–899.

Fraleigh, J. B. (2014). *A first course in abstract algebra*. Harlow, Essex: Pearson.

Fukawa-Connelly, T. (2012). A case study of one instructor's lecture-based teaching of proof in abstract algebra: Making sense of her pedagogical moves. *Educational Studies in Mathematics*, *81*, 325–345.

Fukawa-Connelly, T. (2012). Classroom sociomathematical norms for proof presentation in undergraduate abstract algebra. *Journal of Mathematical Behavior*, *31*, 401–416.

Fukawa-Connelly, T. (2015). Responsibility for proving and defining in abstract algebra class. *International Journal of Mathematical Education in Science and Technology*, *47*, 733–749.

Fukawa-Connelly, T., Weber, K., & Mejía-Ramos, J. P. (2017). Informal content and student note-taking in advanced mathematics classes. *Journal for Research in Mathematics Education*, *48*, 567–579.

Fukawa-Connelly, T. P., & Newton, C. (2014). Analyzing the teaching of advanced mathematics courses via the enacted example space. *Educational Studies in Mathematics*, *87*, 323–349.

Giaquinto, M. (2007). *Visual thinking in mathematics*. Oxford: Oxford University Press.

Goldenberg, P., & Mason, J. (2008). Shedding light on and with example spaces. *Educational Studies in Mathematics*, *69*, 183–194.

Green, J. A. (1965). *Sets & groups: A first course in algebra*. London: Routledge & Kegan Paul.

Hadamard, J. (1945). *The psychology of invention in the mathematical field* (2nd ed.). New York: Dover Publications.

Hadar, N., & Hadass, R. (1981). Between associativity and commutativity. *International Journal of Mathematical Education in Science and Technology*, *12*, 535–539.

Harel, G., & Tall, D. (1989). The general, the abstract, and the generic in advanced mathematics. *For the Learning of Mathematics*, *11*(1), 38–42.

Hausberger, T. (2017). The (homo)morphism concept: Didactic transposition, meta-discourse and thematisation. *International Journal of Research in Undergraduate Mathematics Education*, *3*, 417–443.

Hayward, C. N., Kogan, M., & Laursen, S. L. (2016). Facilitating instructor adoption of inquiry-based learning in college mathematics. *International Journal of Research in Undergraduate Mathematics Education*, *2*, 59–82.

Hazzan, O. (1994). A students' belief about the solutions of the equation $x = x^{-1}$ in a group. In J. P. da Ponte & J. F. Matos (Eds.), *Proceedings of the 18th international conference on the psychology of mathematics education* (Vol. 3, pp. 49–56). Lisbon, Portugal: IGPME.

Hazzan, O. (1999). Reducing abstraction level when learning abstract algebra concepts. *Educational Studies in Mathematics*, *40*, 71–90.

Hazzan, O. (2001). Reducing abstraction: The case of constructing an operation table for a group. *Journal of Mathematical Behavior*, *20*, 163–172.

Hazzan, O., & Leron, U. (1996). Students' use and misuse of mathematical theorems: The case of Lagrange's theorem. *For the Learning of Mathematics*, *16*(1), 23–26.

Hodds, M., Alcock, L., & Inglis, M. (2014). Self-explanation training improves proof comprehension. *Journal for Research in Mathematics Education*, *45*, 62–101.

Hoyles, C., & Küchemann, D. (2002). Students' understanding of logical implication. *Educational Studies in Mathematics*, *51*, 193–223.

Hub, A., & Dawkins, P. C. (2018). On the construction of set-based meanings for the truth of mathematical conditionals. *Journal of Mathematical Behavior*, *50*, 90–102.

Inglis, M., & Alcock, L. (2012). Expert and novice approaches to reading mathematical proofs. *Journal for Research in Mathematics Education*, *43*, 358–390.

Inglis, M., & Attridge, N. (2016). *Does mathematical study develop logical thinking? Testing the theory of formal discipline*. London: World Scientific.

Inglis, M., & Mejía-Ramos, J. P. (2009). On the persuasiveness of visual arguments in mathematics. *Foundations of Science*, *14*, 97–110.

Inglis, M., Mejia-Ramos, J. P., Weber, K., & Alcock, L. (2013). On mathematicians' different standards when evaluating elementary proofs. *Topics in Cognitive Science*, *5*, 270–282.

Inglis, M., & Simpson, A. (2008). Conditional inference and advanced mathematical study. *Educational Studies in Mathematics*, *67*, 187–204.

Inglis, M., & Simpson, A. (2009). Conditional inference and advanced mathematical study: Further evidence. *Educational Studies in Mathematics*, *72*, 185–198.

Johnson, E., Keller, R., & Fukawa-Connelly, T. (2018). Results from a survey of abstract algebra instructors across the United States: Understanding the choice to (not) lecture. *International Journal of Research in Undergraduate Mathematics Education*, *4*, 254–285.

Kapler, I. V., Weston, T., & Wiseheart, M. (2015). Spacing in a simulated undergraduate classroom: Long-term benefits for factual and higher-level learning. *Learning and Instruction*, *36*, 38–45.

Karpicke, J. D., & Blunt, J. R. (2011). Retrieval practice produces more learning than elaborative studying with concept mapping. *Science*, *331*(6018), 772–775.

Kirshner, D., & Awtry, T. (2004). Visual salience of algebraic transformations. *Journal for Research in Mathematics Education*, *35*, 224–257.

Kleiner, I. (1986). The evolution of group theory: A brief survey. *Mathematics Magazine*, *59*, 195–215.

Kleiner, I. (1999). Field theory: From equations to axiomatization. *American Mathematical Monthly*, *10*, 677–684.

Knuth, E., Zaslavsky, O., & Ellis, A. (2017). The role and use of examples in learning to prove. *Journal of Mathematical Behavior*, *53*.

Koriat, A., Bjork, R. A., Sheffer, L., & Bar, S. K. (2004). Predicting one's own forgetting: The role of experience-based and theory-based processes. *Journal of Experimental Psychology: General*, *133*(4), 643.

Kornell, N. (2009). Optimising learning using flashcards: Spacing is more effective than cramming. *Applied Cognitive Psychology*, *23*(9), 1297–1317.

Krupnik, V., Fukawa-Connelly, T., & Weber, K. (2018). Students' epistemological frames and their interpretation of lectures in advanced mathematics. *Journal of Mathematical Behavior*, *49*, 173–183.

Lai, Y., & Weber, K. (2014). Factors mathematicians profess to consider when presenting pedagogical proofs. *Educational Studies in Mathematics*, *85*, 93–108.

Lai, Y., Weber, K., & Mejía-Ramos, J.-P. (2012). Mathematicians' perspectives on features of a good pedagogical proof. *Cognition and Instruction, 30*, 146–169.

Lajoie, C., & Mura, R. (2000). What's in a name? A learning difficulty in connection with cyclic groups. *For the Learning of Mathematics, 20*(3), 29–33.

Larsen, S. (2009). Reinventing the concepts of group and isomorhpism: The case of Jessica and Sandra. *Journal of Mathematical Behavior, 28*, 119–137.

Larsen, S. (2010). Struggling to disentangle the associative and commutative properties. *For the Learning of Mathematics, 30*(1), 38–43.

Larsen, S. (2013). A local instructional theory for the guided reinvention of the group and isomorphism concepts. *Journal of Mathematical Behavior, 32*, 712–725.

Larsen, S., Johnson, E., & Bartlo, J. (2013). Designing and scaling up an innovation in abstract algebra. *The Journal of Mathematical Behavior, 32*(4), 693–711.

Lem, S., Onghena, P., Verschaffel, L., & Van Dooren, W. (2017). Using refutational text in mathematics education. *ZDM Mathematics Education, 49*, 509–518.

Leron, U., & Dubinsky, E. (1995). An abstract algebra story. *American Mathematical Monthly, 102*, 227–242.

Leron, U., Hazzan, O., & Zazkis, R. (1995). Learning group isomorphism: A crossroads of many concepts. *Educational Studies in Mathematics, 29*, '153–174.

Lew, K., Fukawa-Connelly, T. P., Mejía-Ramos, J. P., & Weber, K. (2016). Lectures in advanced mathematics: Why students might not understand what the mathematics professor is trying to convey. *Journal for Research in Mathematics Education, 47*, 162–198.

Lew, K., & Meía-Ramos, J. P. (2019). Linguistic conventions of mathematical proof writing at the undergraduate level: Mathematicians' and students' perspectives. *Journal for Research in Mathematics Education, 50*, 121–155.

Lew, K., & Zazkis, D. (2019). Undergraduate mathematics students' at-home exploration of a prove-or-disprove task. *Journal of Mathematical Behavior*, 100674.

Lockwood, E., Ellis, E. B., & Lynch, A. G. (2016). Mathematicians' example-related activity when exploring and proving conjectures. *International Journal of Research in Undergraduate Mathematics Education, 2*, 165–196.

Mason, J., & Pimm, D. (1984). Generic examples: Seeing the general in the particular. *Educational Studies in Mathematics*, *15*, 277–289.

Mejía-Ramos, J.-P., Fuller, E., Weber, K., Rhoads, K., & Samkoff, A. (2012). An assessment model for proof comprehension in undergraduate mathematics. *Educational Studies in Mathematics*, *79*, 3–18.

Mejía-Ramos, J.-P., & Weber, K. (2014). Why and how mathematicians read proofs: Further evidence from a survey study. *Educational Studies in Mathematics*, *85*, 161–173.

Melhuish, K. (2018). Three conceptual replication studies in group theory. *Journal for Research in Mathematics Education*, *49*, 9–38.

Melhuish, K., Larsen, S., & Cook, S. (2019). When students prove a theorem without explicitly using a necessary condition: Digging into a subtle problem from practice. *International Journal of Research in Undergraduate Mathematics Education*, *5*, 205–227.

Melhuish, K. M. (2019). The group theory concept assessment: A tool for measuring conceptual understanding in introductory group theory. *International Journal of Research in Undergraduate Mathematics Education*, *5*, 359–393.

Melhuish, K. M., & Fagan, J. B. (2017). Exploring student conceptions of binary operation. In *Proceedings of the twentieth annual conference on research in undergraduate mathematics education*. San Diego, CA: RUME.

Michener, E. R. (1978). Understanding understanding mathematics. *Cognitive Science*, *2*, 361–383.

Mills, M. (2014). A framework for example usage in proof presentations. *Journal of Mathematical Behavior*, *33*, 106–118.

Moore, R. (1994). Making the transition to formal proof. *Educational Studies in Mathematics*, *27*, 249–266.

Moore, R. C. (2016). Mathematics professors' evaluation of students' proofs: A complex teaching practice. *International Journal of Research in Undergraduate Mathematics Education*, *2*, 246–278.

Nardi, E. (2000). Mathematics undergraduates' responses to semantic abbreviations, 'geometric' images and multi-level abstractions in group theory. *Educational Studies in Mathematics*, *43*, 169–189.

Nicholson, J. (1993). The development and understanding of the concept of quotient group. *Historia Mathematica*, *20*(1), 68–88.

Novotná, J., & Hoch, M. (2008). How structure sense for algebraic expressions or equations is related to structure sense for abstract algebra. *Mathematics Education Research Journal*, *20*, 93–104.

Ott, N., Brünken, R., Vogel, M., & Malone, S. (2018). Multiple symbolic representations: The combination of formula and text supports problem solving in the mathematical field of propositional logic. *Learning and Instruction*, *58*, 88–105.

Panse, A., Alcock, L., & Inglis, M. (2018). Reading proofs for validation and comprehension: An expert-novice eye-movement study. *International Journal of Research in Undergraduate Mathematics Education*, *4*, 357–375.

Peled, I., & Zaslavsky, O. (1997). Counter-examples that (only) prove and counter-examples that (also) explain. *Focus on Learning Problems in Mathematics*, *19*, 49–61.

Perry, W. G. (1988). Different worlds in the same classroom. In P. Ramsden (Ed.), *Improving learning: New perspectives* (pp. 145–161). London: Kogan Page.

Pinter, C. C. (1982). *A book of abstract algebra*. Mineola, NY: Dover Publications, Inc.

Pinto, A. (2019). Variability in the formal and informal content instructors convey in lectures. *Journal of Mathematical Behavior*, *54*, 100680.

Pinto, A., & Karsenty, R. (2018). From course design to presentations of proofs: How mathematics professors attend to student independent proof reading. *Journal of Mathematical Behavior*, *49*, 129–144.

Pritchard, D. (2010). Where learning starts? A framework for thinking about lectures in university mathematics. *International Journal of Mathematical Education in Science and Technology*, *41*, 609–623.

Raman, M. (2004). Epistemological messages conveyed by three high-school and college mathematics textbooks. *Journal of Mathematical Behavior*, *23*, 389–404.

Rittle-Johnson, B., Loehr, A. M., & Durkin, K. (2017). Promoting self-explanation to improve mathematics learning: A meta-analysis and instructional design principles. *ZDM Mathematics Education*, *49*, 599–611.

Rohrer, D., Dedrick, R. F., & Stershic, S. (2015). Interleaved practice improves mathematics learning. *Journal of Educational Psychology*, *107*, 900–908.

Rohrer, D., & Pashler, H. (2010). Recent research on human learning challenges conventional instructional strategies. *Educational Researcher*, *39*, 406–412.

Roy, M., & Chi, M. T. H. (2005). The self-explanation principle in multimedia learning. In E. Mayer (Ed.), *The Cambridge handbook of multimedia learning* (pp. 271–286). Cambridge: Cambridge University Press.

Roy, S., Alcock, L., & Inglis, M. (2017). Multimedia resources designed to support learning from written proofs: An eye-movement study. *Educational Studies in Mathematics, 96*, 249–266.

Samkoff, A., Lai, Y., & Weber, K. (2012). On the different ways that mathematicians use diagrams in proof construction. *Research in Mathematics Education, 14*(1), 49–67.

Sandefur, J., Mason, J., Stylianides, G. J., & Watson, A. (2013). Generating and using examples in the proving process. *Educational Studies in Mathematics, 83*, 323–340.

Savic, M. (2015). The incubation effect: How mathematicians recover from proving impasses. *The Journal of Mathematical Behavior, 39*, 67–78.

Schoenfeld, A. H. (1985). *Mathematical problem solving.* San Diego: Academic Press.

Schotter, E. R., Tran, R., & Rayner, K. (2014). Don't believe what you read (only once): Comprehension is supported by regressions during reading. *Psychological Science, 25*, 1218–1226.

Segal, J. (2000). Learning about mathematical proof: Conviction and validity. *Journal of Mathematical Behavior, 18*(2), 191–210.

Selden, A., & Selden, J. (1987). Errors and misconceptions in college level theorem proving. In *Proceedings of the second international seminar on misconceptions and educational strategies in science and mathematics* (pp. 457–470). New York: Cornell University.

Selden, A., & Selden, J. (2003). Validations of proofs considered as texts: Can undergraduates tell whether an argument proves a theorem? *Journal for Research in Mathematics Education, 34*, 4–36.

Selden, J., & Selden, A. (1995). Unpacking the logic of mathematical statements. *Educational Studies in Mathematics, 29*, 123–151.

Sfard, A. (1991). On the dual nature of mathematical conceptions: Reflections on processes and objects as different sides of the same coin. *Educational Studies in Mathematics, 22*, 1–36.

Sfard, A., & Linchevski, L. (1994). The gains and pitfalls of reification – the case of algebra. *Educational Studies in Mathematics, 26*, 191–228.

Shepherd, M. D., Selden, A., & Selden, J. (2012). University students' reading of their first-year mathematics textbooks. *Mathematical Thinking and Learning*, *14*, 226–256.

Shepherd, M. D., & van de Sande, C. C. (2014). Reading mathematics for understanding—from novice to expert. *The Journal of Mathematical Behavior*, *35*, 74–86.

Simpson, A., & Stehlíková, N. (2006). Apprehending mathematical structure: A case study of coming to understand a commutative ring. *Educational Studies in Mathematics*, *61*, 347–371.

Skemp, R. R. (1976). Relational understanding and instrumental understanding. *Mathematics Teaching*, *77*, 20–26.

Stylianides, A. J., & Stylianides, G. J. (2009). Proof constructions and evaluations. *Educational Studies in Mathematics*, *72*, 237–253.

Stylianides, A. J., Stylianides, G. J., & Philippou, G. N. (2004). Undergraduate students' understanding of the contraposition equivalence rule in symbolic and verbal contexts. *Educational Studies in Mathematics*, *55*, 133–162.

Stylianou, D. A., & Silver, E. A. (2004). The role of visual representations in advanced mathematical problem solving: An examination of expert-novice similarities and differences. *Mathematical Thinking and Learning*, *6*, 353–387.

Tall, D. (2008). The transition to formal thinking in mathematics. *Mathematics Education Research Journal*, *20*(2), 5–24.

Tall, D. O. (1995). Cognitive development, representations and proof. In *Proceedings of justifying and proving in school mathematics* (pp. 27–38). London: IoE.

Tirosh, D., Hadass, R., & Movshovitz-Hadar, N. (1991). Overcoming overgeneralizations: The case of commutativity and associativity. In F. Furinghetti (Ed.), *Proceedings of the 15th annual conference of the international group for the psychology of mathematics education* (Vol. 3, pp. 310–315). Assisi, Italy: IGPME.

Usiskin, Z. (1975a). Applications of groups and isomorphic groups to topics in the standard curriculum, grades 9-11: Part I. *The Mathematics Teacher*, *68*, 99–106.

Usiskin, Z. (1975b). Applications of groups and isomorphic groups to topics in the standard curriculum, grades 9-11: Part II. *The Mathematics Teacher*, *68*, 235–246.

Vermetten, Y. J., Vermunt, J. D., & Lodewijks, H. G. (2002). Powerful learning environments? How university students differ in their response to instructional measures. *Learning and Instruction*, *12*, 263–284.

Vinner, S. (1991). The role of definitions in teaching and learning. In D. O. Tall (Ed.), *Advanced mathematical thinking* (pp. 65–81). Dordrecht: Kluwer.

Wasserman, N. H. (2014). Introducing algebraic structures through solving equations: Vertical content knowledge for mathematics teachers. *PRIMUS*, *24*, 191–214.

Wasserman, N. H. (2016). Abstract algebra for algebra teaching: Influencing school mathematics instruction. *Canadian Journal of Science, Mathematics and Technology Education*, *16*, 28–47.

Weber, K. (2001). Student difficulty in constructing proofs: The need for strategic knowledge. *Educational Studies in Mathematics*, *48*, 101–119.

Weber, K. (2008). How mathematicians determine if an argument is a valid proof. *Journal for Research in Mathematics Education*, *39*, 431–459.

Weber, K. (2009). How syntactic reasoners can develop understanding, evaluate conjectures, and generate examples in advanced mathematics. *Journal of Mathematical Behavior*, *28*, 200–208.

Weber, K. (2010). Mathematics majors' perceptions of conviction, validity and proof. *Mathematical Thinking and Learning*, *12*, 306–336.

Weber, K. (2012). Mathematicians' perspectives on their pedagogical practices with respect to proof. *International Journal of Mathematical Education in Science and Technology*, *43*, 463–482.

Weber, K. (2015). Effective proof reading strategies for comprehending mathematical proofs. *International Journal of Research in Undergraduate Mathematics Education*, *1*, 289–314.

Weber, K., & Alcock, L. (2004). Semantic and syntactic proof productions. *Educational Studies in Mathematics*, *56*, 209–234.

Weber, K., & Alcock, L. (2005). Using warranted implications to understand and validate proofs. *For the Learning of Mathematics*, *25*(1), 34–38.

Weber, K., Inglis, M., & Mejía-Ramos, J.-P. (2014). How mathematicians obtain conviction: Implications for mathematics instruction and research on epistemic cognition. *Educational Psychologist*, *49*, 36–58.

Weber, K., & Larsen, S. (2008). Teaching and learning abstract algebra. In M. Carlson & C. Rasmussen (Eds.), *Making the connection: Research and teaching in undergraduate mathematics* (pp. 139–152). Washington, DC: MAA.

Weber, K., & Mejí-Ramos, J.-P. (2011). Why and how mathematicians read proofs: An exploratory study. *Educational Studies in Mathematics, 76,* 329–344.

Weber, K., & Mejía-Ramos, J. P. (2014). Mathematics majors' beliefs about proof reading. *International Journal of Mathematical Education in Science and Technology, 45*(1), 89–103.

Weinberg, A., Wiesner, E., Benesh, B., & Boester, T. (2012). Undergraduate students' self-reported use of mathematics textbooks. *PRIMUS, 22,* 152–175.

Weinberg, A., Wiesner, E., & Fukawa-Connelly, T. (2014). Students' sense-making frames in mathematics lectures. *Journal of Mathematical Behavior, 33,* 168–179.

Whitelaw, T. A. (1978). *An introduction to abstract algebra.* Glasgow: Blackie.

Wong, R. M. F., Lawson, M. J., & Keeves, J. (2002). The effects of self-explanation training on students' problem solving in high-school mathematics. *Learning and Instruction, 12,* 233–262.

Zandieh, M., Larsen, S., & Nunley, D. (2008). Proving starting from informal notions of symmetry and transformations. In M. Carlson & C. Rasmussen (Eds.), *Making the connection: Research and teaching in undergraduate mathematics* (pp. 275–287). Washington, DC: MAA.

Zaslavsky, O., & Peled, I. (1996). Inhibiting factors in generating examples by mathematics teachers and student teachers: The case of binary operation. *Journal for Research in Mathematics Education, 27,* 67–78.

Zazkis, D., Weber, K., & Mejía-Ramos, J. P. (2015). Two proving strategies of highly successful mathematics majors. *The Journal of Mathematical Behavior, 39,* 11–27.

Zazkis, R., & Dubinsky, E. (1996). Dihedral groups: A tale of two interpretations. In J. Kaput, A. H. Schoenfeld, & E. Dubinsky (Eds.), *Research in collegiate mathematics education II* (pp. 61–82). Providence, RI: American Mathematical Society.

Zazkis, R., Dubinsky, E., & Dautermann, J. (1996). Coordinating visual and analytic strategies: A study of students' understanding of the group D_4. *Journal for Research in Mathematics Education, 27,* 435–457.

Zhen, B., Weber, K., & Mejía-Ramos, J.-P. (2016). Mathematics majors' perceptions of the admissibility of graphical inferences in proofs. *International Journal of Research in Undergraduate Mathematics Education, 2,* 1–29.

INDEX

phi xiv, 185
physical object 3, 10
plane 49, 94, 101, 125, 126, 198, 199
polar 125
polynomial xiv, 226, 227
precision 38
premise 31, 54, 55, 61, 63, 192, 196
prime 114, 121, 177, 179, 217,
    218, 233
problem solving 33, 34, 61
processing 70, 71
professor xvi
projection 95
proof
  chain of deductions 32
  construct 35, 60–65, 192, 196,
    218, 261
  conviction 47
  formal 61–63
  premise to conclusion 63
  problem solving 61
  reading 49, 58, 60
  reconstruct 33, 42
  relate to examples 53
  short 32, 53
  strategy 61
  study 32, 33
  uniqueness 52, 53
  writing 60, 62–65, 194
proper subgroup 114
psi xiv, 185
psychology 68
pure 7, 68, 197, 259

$(\mathbb{Q}, +)$ 248
$\mathbb{Q}$ xiii, 17, 46, 123, 222
quantifier
  existential 51, 52
  for all 18, 21, 51, 110
  function addition 91, 92
  order 22, 52
  there exists 21, 51
  universal 51–53, 223, 243
quaternions 235

quotient
  group 99, 153–155, 157, 158, 162,
    164, 166, 167, 169, 181, 212, 216,
    244, 246, 248
  ring 244, 245, 247–249, 251, 252, 255

$(\mathbb{R} \backslash \{0\}, \times)$ 203
$\mathbb{R}$ xiii, 17, 123, 124, 222
$\mathbb{R} \times \mathbb{R}$ 94, 101
$\mathbb{R}^2$ 94, 101
rational number xiii, 17, 35, 123, 203
reading v, 14–17, 49, 56, 57, 60, 65, 74
real number xiii, 17, 21, 36, 123, 124
recall 72
recognize 72
reconstruction 71, 75, 231
rectangle 135, 198
reflection
  composition 128, 129
  dihedral group 129, 133, 134
  matrix 95
  permutation notation 139
  plane 95
  square 135
  subgroup 128, 132
  triangle 9, 96, 151
reflexive 46
refute 53
relation
  $<$ on $\mathbb{Q}$ 46
  $\equiv$ (mod 12) on $\mathbb{Z}$ 46
  and generator 116, 131, 134
  dihedral 131, 134
  equivalence 45, 46, 85
  is isomorphic to 189
  not an operation 46
  notation xiii, 46
  symmetric 189
relatively prime 113, 114, 233
remainder 41, 42, 46, 85, 154, 174, 212,
    224, 246–248
remembering 70
representation
  circle 41, 121, 122, 125, 126